D1486399

The Plumbat Affair

Additional research by Leni Gillman

Elaine Davenport,
Paul Eddy and Peter Gillman

THE PLUMBAT AFFAIR

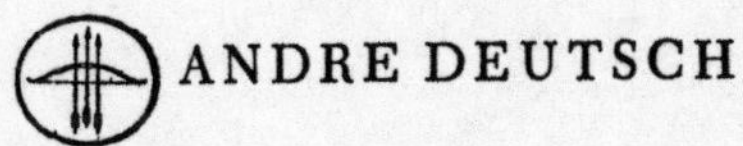

First published 1978 by
André Deutsch Limited
105 Great Russell Street London WC1

Printed in Great Britain by
Tonbridge Printers Ltd
Tonbridge, Kent

ISBN 0 233 97016 9

Contents

List of Plates

Authors' Note

The Plumbat Affair is based primarily on inquiries we made in seventeen countries and some 200 interviews. We conducted most of the interviews ourselves but we owe a considerable debt to a number of other journalists who either gave us the product of their own interviews or questioned witnesses on our behalf: in Paris, Antony Terry, European Editor of *The Sunday Times*; in Hamburg, Jay Tuck; in Mainz, Gunter Kitzinger; in Milan, Dalbert Hallenstein; in Oslo, Arvid Bryne; and in Copenhagen, Ole Schierbeck. We also owe thanks to Tony Geraghty of *The Sunday Times* whose original investigation of the disappearance of the uranium provided valuable guidance.

We benefited, too, from the work of authors who have written about related topics. A full bibliography appears on pages 183–84.

Finally, we have some personal debts to pay. Georgetta Moliterno, Laurie Zimmerman and Tammy Pittman, all in New York, and June Pratt in London did dozens of jobs for us patiently and well. Shirley Poluck typed the manuscript with remarkable energy and Gordon Phillips compiled the index. Robert Ducas, our agent and friend, provided, as always, magnificent encouragement. And, most of all, we have to thank Leni Gillman whose research and administration made this book possible.

Elaine Davenport
Paul Eddy
Peter Gillman

February 1978

Cast of Characters

The reader is invited to use this list of main characters both as a preview of what is to come and, later, as a reference for a name forgotten.

MAGNHILD AANESTAD Norwegian detective with slender wrists and a liking for simple summer dresses

DAN AERBEL Mossad agent who gave away every secret he knew, and . . .

DEBRA AERBEL his Israeli-born wife

FELIX AMIOT built the Israeli gunboats in Cherbourg

MEIR AMIT replaced Isser Harel as director of Mossad

ASMARA CHEMIE of Wiesbaden bought 200 tons of uranium oxide

ROLF BAEHR Mossad agent caught up in murder

WILHELM BARGON Asmara Chemie salesman who broke down the resistance of front-office secretaries

PETER BARROW temporary skipper of the *Scheersberg A* with a predilection for dark blue shirts and trousers

KEMAL BENAMENE perhaps a courier for Black September, he led Mossad to Lillehammer

DAVID BEN-GURION Israel's first Prime Minister who ordered Mossad to employ restraint

JUERGEN-ADOLF BINDER a senior partner of August Bolten, he looked uneasily on the sale of the *Scheersberg A*

BISCAYNE TRADERS SHIPPING CORPORATION gave the *Scheersberg A* a Liberian flag and a new and mysterious crew

BKA the West German equivalent of the FBI, conducted an energetic investigation

BLACK SEPTEMBER planned the Munich massacre and caused Mossad to seek revenge

BND West Germany's CIA, helped plant an Israeli agent in Cairo

AUGUST BOLTEN a Hamburg shipping company which sold the *Scheersberg A* to the elusive Burham Yarisal

PIERRE BOMMELLE Euratom's bloodhound who took his Geiger counter to West Germany in search of the uranium

AHMED BOUCHIKI met a stranger in Lillehammer and marked himself for death, and . . .

TORILL BOUCHIKI his Norwegian wife

MILA BRENNER set up Starboat in Panama to get the gunboats out of Cherbourg

GUIDO BRUNNER became Energy Commissioner of the EEC and tried to explain the Plumbat cover-up

CHIMAGAR the Moroccan company originally selected to receive the uranium

CIEEMG gave French government approval for the Starboat deal

CMN Felix Amiot's company which was Cherbourg's lifeblood

FRANCISCO COUSILLAS replaced Peter Barrow and wanted to know where the *Scheersberg A* had been

RAOUL COUSIN another member of Mossad's 'hit team'

MOSHE DAYAN Israel's Minister of Defence, he argued for conversion of the nuclear option

DENIS DEWEZ sold Asmara Chemie its uranium and then, innocently, nearly wrecked the plan

MICHAEL DEWITT employed Aerbel as a salesman in Europe

DIMONA the site of Israel's nuclear reactor, guarded mercilessly

MICHAEL DORF was arrested for the Lillehammer murder but later cleared

ALDO EGGERS-LURA was sorry when Aerbel gave up his job as a salesman of Danish furniture

E-GRUPPA the Norwegian police squad which investigated the Lillehammer murder

DAN ERT/ERTESCHICK/ERTZ/ERTL the different names Aerbel used

EURATOM Europe's nuclear watchdog which had no bite and little bark

EUROPEAN COMMISSION the EEC's executive which decided to cover up the Plumbat affair

EEC the European Economic Community, then made up of Belgium, France, Italy, Luxembourg, the Netherlands and West Germany

EUROPEAN PARLIAMENT was furious when it found out about the cover-up

COMMANDER EZRA obeyed orders and took the gunboats from Cherbourg to Haifa

JOB FASS made knick-knacks which Aerbel sold to Galeries Lafayette

GEORGES FRANJISTAS bought the *Scheersberg A* from Aerbel and renamed it after his daughter

CHARLES de GAULLE the French President who said 'non' when Israel wanted more aeroplanes and uranium

REINHARD GEHLEN ran the West German BND

ABRAHAM GEHMER another Mossad agent, he pretended to be a schoolteacher from England

MARIANNE GLADNIKOFF short and slightly plump she found she had been recruited for a Mossad 'hit team'

GUNNAR HAARSTAD head of Norway's secret service he realized the enormous implications of the secret in the Lillehammer file

WILHELM HAFERKAMP was EEC Energy Commissioner when Mossad mounted Operation Plumbat

LUDWIG HANSEN was ordered to abandon his post as captain of the *Scheersberg A*

ISSER HAREL Mossad's first director, he captured Adolf Eichmann but then went too far

HAROULA became the *Scheersberg A*'s new name when Georges Franjistas bought her

NORA HEFFNER another member of Mossad's 'hit team'

MARCEL HEYNEN worked for Ziegler & Co and drank whisky with Captain Barrow

TORBEN HVIID lived on the edge of the law and gave Aerbel free admission to 'Love In'

JONATHAN INGLEBY Mossad agent and killer

ENRICO JACCHIA led Euratom's investigation

ROLF JAHRMANN head of E-Gruppa, he did not understand the significance of Aerbel's explosive confession

PETER KOERNER returned to the *Scheersberg A* and found that two vital pages from the logbook had been torn out

GERARD LAFOND unwittingly provided cover for another member of the Mossad 'hit team'

GERD LANZ set up Biscayne Traders Shipping Corporation but refused to reveal the identity of his client

HERBERT LETTKO Asmara Chemie's persistent salesman, he missed the point of a joke

PAUL LEVENTHAL heard intoxicating rumours on Capitol Hill and broke the Plumbat story

LEIF LIER heard an amazing confession from Marianne Gladnikoff

MORDECAI LIMON ordered the gunboats to sail from Cherbourg

WOLFGANG LOTZ spied for Israel in Cairo and joined in an explosive campaign to deter German scientists

KAJ LUND used his legal skills to set up 'Love In' and dissolve Biscayne Traders

RUDOLF MADSEN was taken aback when Aerbel sold souvenirs in Libya

GEORG MANNER led Mossad's hit team in Lillehammer

GOLDA MEIR Israel's Prime Minister, she ordered revenge for the Munich massacre

METALLURGIE HOBOKEN-OVERPELT stored the uranium near the red-brick village with an improbable legend

JEAN MIKOLAJCAK was titular head of SGM's uranium division and stressed that the deal was confidential

UWE MOELLER helped Yarisal to buy the *Scheersberg A* and Aerbel to sell it

MOSSAD the Institute was the most battle-hardened secret service in the world

YUVAL NE'EMAN computerized Mossad's files and designed Israel's bomb

FELIX OBOUSSIER gave Asmara Chemie the approval it needed

LESLIE ORBAUM was the alias used by Abraham Gehmer

KNUD PEDERSEN provided the Cherbourg connection and died mysteriously

SHIMON PERES Israeli cabinet minister, he rented the

smallest car he could find to drive to West Germany

GUSTAV PISTAUER told Aerbel to stop asking questions

GEORGES POMPIDOU President of France, he was reckoned by Mossad to be no de Gaulle

PFLP Popular Front for the Liberation of Palestine which struck at Israel through terrorism

SYLVIA RAPHAEL cruised on a yacht with Aerbel and joined the Lillehammer 'hit team'

STEINAR RAVLO knew that Aerbel was going to confess

KLAUS REHWAGEN hired Ziegler & Co to get the uranium to Antwerp

PATRICIA ROXBURGH Sylvia Raphael's alias

SAICA the Italian paint company which ran into a little trouble

ALI HASSAN SALAMEH planned the Munich massacre and became Mossad's prime target

HERBERT SCHARF named his new company after a town in Ethiopia

SCHEERSBERG A showed her age but kept her silent rendezvous in the Mediterranean

OLE SCHIERBECK remembered a meeting in a Hamburg hotel and revealed the Cherbourg connection

HERBERT SCHULZEN was called 'the Nazi pilot' but his name went into Mossad's files

FRANCESCO SERTORIO was not bothered that his paint company had no idea how to treat uranium

JEAN-LUC SEVENIER served with Aerbel in Mossad's advance team

SGM received an order for uranium which it could easily fill

MARTIN SIEM impressed the French with his entry in the international edition of *Who's Who*

STARBOAT said it wanted to use gunboats for oil exploration

ZWI STEINBERG Mossad agent with a double identity – 'My God,' said the policeman, 'that's Zipstein'

TAMARA Manner's mistress and a killer

FIRST OFFICER TILNEY nearly steered the *Scheersberg A* into trouble

GEOFFREY TOLLMAN was not sorry when his partnership with Aerbel ended

INGO TREPEL had a bright idea which got Schulzen started

JOSEF VERHULST got up early to make sure the uranium was loaded

VIKING had no substance but reduced Aerbel's unwelcome air of mystery

HAKON WIKER Norwegian state prosecutor, he found himself in the hot seat

BURHAM YARISAL founded a company and bought a ship and disappeared when it suited him

ZWI ZAMIR replaced Amit as head of Mossad and changed the rules – with disastrous consequences

ZIEGLER & COMPANY made all the arrangements and paid the bill for the consequences of an unfortunate accident

VICTOR ZIPSTEIN Steinberg's alias when he drove Admiral Limon's car

Chapter 1
Murder in Norway

On a Saturday evening in July 1973 Dan Aerbel was standing in the main street of Lillehammer, a placid Norwegian resort town 80 miles north of Oslo. On the other side of the street was Lillehammer's only cinema, showing the World War II adventure epic *Where Eagles Dare*. Aerbel was not interested in the movie, but in a man he had watched go into the cinema for the last programme of the day. Aerbel and nine companions now in different parts of Lillehammer were waiting for the movie to end and the man to come out. When he did, they would kill him.

They had been tailing their target around Lillehammer all day. They first sighted him that morning as he relaxed over a cup of coffee at a street cafe in the town centre. They followed him when he went to swim at the Lillehammer pool, and then as he made his way to an apartment block on the edge of town. They staked out the building in four rented cars from which they could cover all the exits. They saw him emerge around 7:30 that evening in the company of a woman wearing a bright yellow raincoat which did not disguise the fact that she was heavily pregnant. They followed the couple as they walked back into town and saw them disappear into the cinema. Aerbel stood watching the cinema's main entrance from across the street, hoping he was not too conspicuous.

Where Eagles Dare came to its bloody but victorious climax at 10:35 pm. The audience spilled out of the cinema into the warm night air: in the midsummer twilight the man and the woman, wearing a yellow raincoat, were easy to spot. As they turned up Lillehammer's main street, Storgaten, heading for the bus stop, Aerbel followed them on foot. He watched as the last bus of the night pulled up and the couple climbed on board. The message was passed on by walkie-talkie radio: the target was coming.

There were half a dozen people on the bus. It stopped once to let off a small girl and then a second time close to the apartment building where the couple had spent the afternoon. They were the only ones to get off. The bus drew away.

The woman noticed a car parked in the road facing her, its parking lights on. 'We thought it was waiting for the bus to pass so that it could turn,' the woman said later. 'Then it moved slowly towards us. It passed by very close to us. Then it stopped.'

Inside the car, a white Mazda, were three men and a woman, alerted to expect the bus by the radio message. As the bus disappeared up the street two of them, a man and the woman, got out. Each held a handgun, a .22 caliber Beretta fitted with a silencer.

The man who was their target saw them approach. He shouted one word: 'No.'

The pair from the car did not reply. They shot the man six times in the stomach. As he slumped to the ground they shot him twice in the head. As he lay prostrate they shot him six more times in the back.

The pregnant woman could not believe what she saw: it was like a scene from a gangster movie. Even the guns seemed unreal. 'They sounded like cap pistols,' she said. 'I saw bright flashes, many of them. It was all over in seconds.'

She fell to the ground beside the dying man. The two killers ignored her, walked back to the white Mazda, got in and drove away. The pregnant woman was still huddled there when a second car, a green Volvo, drew up. The driver looked across briefly at the forlorn tableau. Certain that the man was dying, he picked up his walkie-talkie and said in English: 'They took him. All cars go home.'

Dan Aerbel was waiting in a white Peugeot in the centre of Lillehammer when the signal came. With him were three of the others who had tracked the target to his death. They drove out of Lillehammer and headed south for Oslo. Five miles down the road the Peugeot stopped at a rendezvous point. The rest of the team were already there. Someone asked one of the killers how things had gone. He replied that a job was a job.

The small convoy of cars, all of them rented, continued their journey to Oslo. Despite the apparent ease with which the task had been accomplished, Dan Aerbel, in the back of the Peugeot, was anything but relaxed. He complained of a stomach ache and took swigs from a bottle of whisky, although he professed to be a teetotaler. The young girl sitting beside him seemed just as nervous. Aerbel took her hand. Even he was not sure whether his gesture was intended to comfort or seduce.

Less than twenty-four hours later, Dan Aerbel was back in Lillehammer on Storgaten, the street where he had tracked the target to the bus stop. This time he was the quarry. He had been arrested in Oslo on suspicion of murder and driven to Lillehammer police station. In a first floor office overlooking Storgaten, Aerbel was facing one of Norway's most persuasive police interrogators, Inspector Steinar Ravlo, a member of Norway's serious crimes squad, the E-Gruppa. Ravlo, a man of classic Nordic looks – blond hair and unrelenting blue eyes – was convinced that Dan Aerbel had helped carry out this bizarre execution, the first murder within Lillehammer town limits in forty years. But Ravlo had no idea why.

At the first interrogation Ravlo got only vigorous denials of Aerbel's involvement. But one night in the station cells was enough to weaken Aerbel's resolve. When Ravlo began his second interrogation on 23 July, two days after the killing, he sensed what was about to happen. Aerbel began by repeating his denials. Then came a long silence. 'My palms were sweating,' says Ravlo. 'I was really excited. I knew he was going to crack.'

Aerbel stared through the window, looking down on the desultory Monday morning traffic in Storgaten. The silence persisted: five minutes, ten, fifteen. Finally Aerbel spoke.

'Okay,' he told Ravlo. 'You don't believe anything I say. I will tell you how it is.'

Day after day for the next three weeks Aerbel talked, and

the story he told was astonishing. Aerbel said that he was an agent for Mossad, Israel's secret service, and had been since 1963. Under the cover of being a salesman, and using half a dozen different names, he had carried out missions for Mossad in France, Italy, Scandinavia – even Libya. Aerbel's latest mission had been the murder in Lillehammer. He claimed that he was a member of an official Israeli assassination squad, originally set up under direct orders from Prime Minister Golda Meir.

It was a confession which Inspector Ravlo regarded with the greatest caution. By most western standards Norway's serious crime rate is very low; also, Norwegian policemen rarely have to cope with the international terrorism that has developed from the ruthless politics of the Middle East. But even when Ravlo knew from the wealth of detail Aerbel was providing that his story must be true, he maintained his air of scepticism. It was a ploy to extract more and more information from Aerbel and it met with remarkable success. The more Ravlo appeared unimpressed by the series of confessions, the more Aerbel sought to convince him they were true. The relationship became a near perfect example of the dependency the interrogator seeks to create in his victim. Aerbel came to view Ravlo as a friend he wanted to please. Ravlo encouraged the friendship by granting Aerbel small favours. In his gratitude, Aerbel arranged for the inspector to receive a small present, sent from Israel.

And so the confessions continued. Aerbel talked about his previous missions; he gave away Mossad's top-secret emergency telephone number in Tel Aviv; he named his Mossad controller and his contact man in Oslo; and he gave away his co-conspirators in the assassination plot.

By the middle of August, when Aerbel had been transferred from Lillehammer to a prison at Trondheim close to the Arctic Circle, and Ravlo had to fly 350 miles in each direction to visit him, Aerbel had only one secret left to give. It was the biggest secret of them all. As soon as Aerbel had blurted it out, he himself realized that. He fell into an uncharacteristic silence and refused to tell Ravlo anything more.

Ironically, what Aerbel had said meant nothing to Ravlo. He wrote a brief report of the conversation which he gave to his boss, the head of E-Gruppa, when they met next day in Oslo. He, too, was mystified and slipped Ravlo's note into the bulging Lillehammer file.

But there were others who did know what it meant – to whom Aerbel's fragment of information was the vital missing part in a puzzle that the intelligence services of Europe and America had been poring over for almost five years. They were agents of Norway's secret service, the *Politiets Overvaakningstjeneste*, who every day checked through the Lillehammer file at Oslo police headquarters to keep abreast of the latest revelations about Mossad, and to pass the choicest parts on to the intelligence services of Norway's allies in NATO. Some time around 15 August 1973, the Norwegian agents read Inspector Ravlo's report of his most recent conversation with Aerbel. What Aerbel had told Ravlo was that once, in Mossad's cause, he had owned a ship called the *Scheersberg A*. The Norwegian secret agents made the connection and a mystery of enormous international implications had at last been resolved.

In November 1968 the *Scheersberg A*, a small and shabby cargo boat laden with 560 metal drums, had set sail from the Belgian port of Antwerp, bound for Genoa, Italy. She never arrived. Two weeks after she was due at Genoa she put in instead at the eastern Turkish port of Iskenderum. Her captain, who said his name was Peter Barrow, told the port authorities that the *Scheersberg A* was empty, which was self-evidently true, and that she had come from Naples, which was a lie. A few days later Barrow and his crew had abandoned the *Scheersberg A* – and disappeared.

The voyage of the *Scheersberg A* had greatly preoccupied the security services of Belgium, Italy and West Germany, together with their partners in the European Common Market, and their allies in NATO. They had all investigated what had happened to the *Scheersberg A*'s cargo, and who could have organized the operation which had resulted in the cargo apparently being spirited away. All of them, including the American CIA, drew a blank.

Now, in August 1973 in Oslo central police station, the Norwegian security men found themselves staring at the answers. Dan Aerbel's temporary ownership of the *Scheersberg*

A could only mean that the true destination of her cargo on that mysterious voyage had been a place called Dimona. And that meant that there could now be no doubt that Israel had acquired the means to develop nuclear weapons.

Chapter 2
A Secret Place, A Secret Service

Dimona. It stands in a parched wilderness of rock and scrub. Its dome-shaped centrepiece might be taken for a basilica; closer inspection shows it to be more like a giant soccer ball. Then there are the slender chimneys, climbing to three times the height of the dome, and the squat buildings clustering round, and the fence laden with warning signs that encircles the whole area. It is a stark intrusion upon its desert surroundings which have remained largely unchanged since Biblical times. It is as though some new race had landed there, and built a colony to survive an alien environment.

It is not easy to get any view of the Dimona Centre. It is only forty miles away from the ancient and bustling town of Beersheba, but the casual visitor heading southwest on the road into the Negev Desert will be intercepted by Israeli patrols before he sights anything more than the tips of Dimona's chimneys. Polite but insistent, the soldiers will turn him back the way he has come. Even if he can avoid the patrols and somehow pass undetected through the invisible electronic barriers that crisscross the desert, he will be halted by the bristling fences that guard Dimona with stern signs ordering him away, warning also that photography is strictly prohibited.

The Centre is guarded mercilessly. In 1967 an Israeli Mirage fighter strayed into the forbidden air space above this place. An Israeli missile shot it down.

Obsessive secrecy shrouded the project from its very beginning. Construction work started in 1958 on a desert site, eight miles from a tiny settlement of Jewish pioneers called Dimona. Israel pretended she was building a textile factory and the nomadic Bedouin tribes roaming their traditional desert territory did not know any different. But as the Dimona settlement, obliged to accommodate more and more workers, grew to the size of a small town, that pretence became harder to maintain. In 1960 an agent of Egypt's secret police, the *Moukhabarat*, reported his suspicions that the spreading industrial complex could scarcely be for the production of synthetic textiles. Egypt passed those suspicions on to the United States.

On 8 December 1960, a United States Air Force photo-reconnaissance plane flew over Dimona Centre. It brought back photographs showing railway lines, high tension wires, chimneys, vast concrete workshops and, most important, the telltale soccer ball dome. The next day experts of the American Central Intelligence Agency told a secret and unannounced meeting of the Joint Congressional Committee on Atomic Energy that Israel was building a nuclear reactor.

The reaction in Washington bordered on panic. Immediately after the Congressional meeting the US Secretary of State Christian Herter summoned the Israeli ambassador Avraham Harman and confronted him with the evidence of the photographs. Herter demanded to know if Israel was planning to build nuclear weapons. The ambassador said he would consult his government.

It was twelve days before he returned with an assurance that the reactor was intended for purely peaceful purposes. On the same day, Israel's Prime Minister David Ben-Gurion stood up in the Knesset, Israel's parliament, and said much the same: Dimona would serve only the needs of 'industry, agriculture, health and science.'

Even in Israel these promises rang hollow. If the reactor were intended solely for the production of energy, why all the secrecy? Why were even members of the Knesset not allowed to visit the site, or examine the budget? And why was Dimona being built by Israel's Ministry of Defence? (This fact had led to the resignation of all the members of the Israel Atomic Energy Commission, with the exception of the chairman.)

These questions went largely unanswered, and construction work at Dimona went on. The reactor was designed to consume uranium, and there lay the cause of the United States' concern. In its natural state uranium consists of two isotopes, U-238 and U-235. The second of these is present only in minute quantities, but it is radioactive and provides the vital ingredient for nuclear weapons. Before it can be used in this way it has to be separated from the U-238. This is an enormously elaborate and costly process of which only advanced industrial nations are capable.

Dimona, however, gave Israel an alternative. In the course of consuming uranium, reactors like Dimona produce Plutonium-239 as a by-product. Provided that the process is halted at the right point, P-239 can be extracted from the spent fuel with relative ease. It is not such a potent substance as U-235, but, nonetheless, it could be used to make a very effective nuclear bomb. The truth was that even if Dimona were initially intended for 'peaceful purposes', it would give Israel the *potential* to build nuclear weapons.

In the fall of 1967 the Israeli Cabinet met in Jerusalem in an air of crisis. Israel itself was still celebrating her crushing victory in the June Six Day War. But the Cabinet had to confront the fact that, however brilliant in military terms, the victory had served only to increase Israel's isolation. Israel had justified her 'preemptive strike' against Syria, Jordan and Egypt on the grounds that France, her main arms supplier, was about to impose an arms embargo. Since the war, France had done precisely that. An enraged General Charles de Gaulle had ordered an immediate halt of arms supplies to Israel and the first casualty had been the fifty Mirage jet fighters which her air force had ordered.

The issue before the Cabinet, however, was not simply how Israel could replace a few aeroplanes. It was rather more basic than that: how to survive. France's decision was but one further indication, in the shifts and vagaries of international politics, that there was not one country she could rely on.

France was by no means the first ally who had turned out to be a fair-weather friend. West Germany had once agreed to supply Israel with arms and then reneged the moment the Arabs found out. Britain, after the Suez fiasco of 1956, had reappraised her Middle East policy on arms supplies, to the benefit of the Arabs, and to Israel's cost. The United States remained steadfast in her policy of supplying Israel with money but not arms, despite all the protests of the American Jewish lobby. Meanwhile, the Soviet Union was resupplying Israel's enemies with most of the arms they could want.

This isolation had long been accepted as a fact of life by Israel's leaders. But what sharpened the debate in the aftermath of the June war was the almost universal hostility shown by Israel's allies to her conquests, and her decision to occupy large areas of Arab territory.

To some extent Israel could compensate for her lost arms supplies by developing her own arms industries, and this, following the 1967 war, she proceeded to do.* But cabinet ministers such as Moshe Dayan and Shimon Peres contended that the burden of acquiring increasing amounts of conventional weapons would eventually become insupportable. They also pointed out that Israel's military strategy had always been based on the belief that the Arab countries could afford to lose a war or two, whereas one Israeli defeat would be her last: If Israel continued to rely on conventional weapons, defeat one day would be inevitable. In short, the cabinet must make the monumental decision which it had so far resisted: to acquire nuclear weapons. Israel had of course given herself the option to do so, by building Dimona. Now Dayan, and others, argued that the time had come to convert that option – to dedicate Dimona to the production of the ultimate deterrent.

It was not an argument some members of the Cabinet, such as Prime Minister Golda Meir, and Yigal Allon, could easily embrace. The moral principles on which the state of Israel was founded do not sit comfortably with a plan to build weapons of mass destruction. But in the end, because of the demonstrable fragility of Israel's international alliances, they accepted the words of Moshe Dayan: Israel had 'no choice.'

*In this endeavour she was greatly helped by a Swiss aircraft engineer named Alfred Frauenknecht. He worked for the Swiss aircraft company, Sulzer Brothers, which manufactured the French Mirage III under licence for the Swiss Air Force. In return for about $200,000 Frauenknecht stole two tons of Mirage blueprints from Sulzer during 1968 and passed them on to a Mossad agent. Frauenknecht was eventually caught and jailed for four and a half years.

The decision was made, but the practical consequences were considerable. Among the problems a nation with military nuclear aspirations has to solve are those of delivery and detonation. First and foremost, however, came the problem of fuel for the reactor. To produce enough Plutonium-239 for one small bomb* would take Dimona almost a year, and consume approaching twenty tons of uranium. To build an effective arsenal, consisting of a dozen such weapons, would take Dimona a decade and would require 200 tons of uranium.

Israel had nothing like that amount of uranium available. When the reactor was completed in 1964 Israel had been supplied with some uranium by France, who had helped her, in secret, to build Dimona in the first place. By 1967 these stocks were fast running out. There was of course no longer any question of France supplying more. Nor would other countries with uranium, or access to it, risk being called Israel's accomplice in acquiring the bomb.

All legitimate channels to uranium were closed. Israel could hardly steal it by force without becoming an international outlaw. Only one solution remained: to acquire uranium by stealth. Late in 1967 the Israeli cabinet gave just that assignment to Israel's central bureau of intelligence and security, Mossad. It was exactly the kind of assignment Mossad relished.

Shortly after the Cabinet's decision, the Israeli Embassy in London gave a small cocktail party to mark the impending return home of its military attache, Brigadier General Zwi Zamir. When someone at the party asked Zamir what his plans were, he replied that he was going into the textile business. That was a joke in the best traditions of Mossad. In Israel 'textiles' is often a euphemism for secret – hence, Dimona was built under the guise of a textile factory. Zamir was hinting that he was returning to Israel to become the new head of Mossad.

Zamir was a former army commander, who had been given charge of his first brigade at the astonishingly young age of

*Small is a relative term: each warhead would have approximately the same explosive force as the atom bomb which killed 100,000 people in Hiroshima on 6 August 1945.

twenty-six. His reputation in action was for checking every detail of a plan before giving it his final approval. His meticulous, circumspect approach was reckoned to be exactly suited to the demands of modern intelligence work, and so, too, was a major aspect of his personality. Zamir was something of an introvert, and he believed that both he, and the agency he was about to take over, should maintain a very low profile. In later years Zamir was to reverse that policy, with disastrous consequences. But at the time of his appointment he was judged by the Israeli cabinet to be the ideal man to take charge of an operation that was more important than any other in Mossad's short but tumultuous history.

The empire Zamir inherited was without doubt the most battle-hardened secret service in the world. Mossad was formed in 1937, when the Jewish settlers in Palestine were fighting on two fronts. On one they faced the Arabs, with whom they were competing for land. On the other were the British, who had governed Palestine under a mandate from the League of Nations since 1923 and were struggling to restrict immigration in order to maintain a balance between Arabs and Jews.

At the same time Europe's Jews were at increasing hazard from the obscene race laws of Nazi Germany and the spreading cancer of anti-Semitism. Senior officers of the Haganah, the underground Jewish army, created an agency dedicated to getting Jews out of Europe and into Palestine. They named it Mossad Lealivah Beth: Mossad means Institute; Lealivah Beth, Immigration route B.

Thousands of Jews escaped from the ghettos of Europe along Mossad's 'route B' with forged papers that Mossad provided. The more difficult part of the operation lay in getting those refugees into Palestine. The Royal Navy set up blockades to prevent illegal immigrants coming in by sea, while the British Army patrolled likely landing places. Mossad manned secret radio posts to help small boats in through the blockades, and met the immigrants on the beaches to lead them to safe houses.

The outbreak of World War II brought Mossad a more ambivalent role. Illegal immigration into Palestine continued, but at the same time Mossad agents trained and fought with the British in the fight against Nazi Germany. Some were trained as parachutists and saboteurs at Britain's Special Operations School near Cairo. They were dropped into occupied eastern Europe to gather intelligence – a mission that was doubly risky for Jews. Later, as the war swung against Hitler, Mossad agents helped foment revolt ahead of the advancing Allied armies. The cost was heavy: of one sizeable group of agents parachuted into Slovakia, for example, only one man survived.

Germany's defeat in 1945 allowed Mossad to resume uncomplicated hostilities with the British. Not only immigrants flowed into Palestine along Mossad's routes but also arms, ammunition, and explosives. Some reached the Jewish terrorist gangs, Irgun and Stern, who harried the British at every turn. Some went to arm the Haganah, increasingly in conflict with the Arabs. Much was stockpiled for the full-scale war with the Arabs which Mossad knew to be inevitable if Israel was to achieve her statehood.

On 14 May 1948, the British, finally weary of trying to keep the two sides apart, pulled out of Palestine. Within hours the Jewish National Council proclaimed the new state of Israel. The Arabs declared war and Egypt, Syria and Jordan attacked simultaneously. As the fighting raged, Mossad was entrusted with the task of supplying arms by every possible means.

Mossad's efforts enabled Israel to resist the Arab invasion, and after three weeks of fighting the United Nations secured a temporary ceasefire. In the four-week lull that followed, Mossad clinched the deal that won Israel the war. The Israel forces had been greatly handicapped by the bewildering variety of arms they had been supplied with. Unit commanders, never knowing what weapons to expect, stocked a dozen types of ammunition. Above all they feared receiving the totally unreliable Italian war surplus which reached them, via scavenging arms dealers, from the scenes of Italian defeats in Ethiopia and Eritrea. Mossad now helped to negotiate the purchase of a vast consignment of arms, supplied by Czechoslovakia with the blessing of the Soviet Union. The ceasefire collapsed on 9 July and the re-armed Israelis attacked on all fronts. They scored a succession of victories from which the

Arab armies never recovered. By January 1949 Israel had won her right to exist.

Mossad's reward for its part in Israel's victory was permanent recognition. David Ben-Gurion, Israel's first Prime Minister, disbanded the host of secret – and at times rival – groups which the long struggle had thrown up.* The new state was equipped with a formal civilian security apparatus and Mossad – the Institute – was placed at its pinnacle. The words Lealivah Beth were dropped and it was made responsible for all intelligence gathering and 'special operations' abroad. It was given offices in the Ha'Kiryah district of north Tel Aviv. Internal security was the province of another agency, Shin Beth, but there was no doubt which was the senior service. Mossad's director, Isser Harel, became chief executive of both, and reported directly to the Prime Minister.

Harel was a veteran of the long Jewish struggle. Born in Russia in 1912, he had emigrated to Palestine at the age of seventeen to work on a kibbutz and was almost immediately recruited into the Jewish resistance. By the time he was appointed head of Mossad he had spent more than half his life in a twilight world of subterfuge, sabotage, and guerrilla action.

Harel's Mossad was, like the man, tough and uncompromising. At first it was preoccupied with revenge, hunting down Nazi war criminals. Its most celebrated victim was Adolf Eichmann, the SS officer who had administered Hitler's final solution. He was traced by Mossad to Argentina, kidnapped and taken back to Israel for trial and execution. That kidnapping was justified in Israel by the enormity of Eichmann's crime, but the government recognized that Mossad could not go on invading friendly countries and openly breaking their laws. In the early 1960's at Prime Minister Ben-Gurion's insistence, Mossad evolved into a more orthodox secret service, concentrating on intelligence gathering and analysis. This more restrained policy did not altogether suit Isser Harel and

*There was Haganah, the main Jewish army; Shai, its intelligence wing; Palmach, an elite commando group that had worked closely with Mossad; Rechesch and Bricha, unaffiliated groups of spies and saboteurs; the Stern and Irgun terrorist gangs and Institute X which coordinated their activities with Haganah.

in 1963, after a furious dispute with Ben-Gurion, which we will come to, he resigned.

His replacement as head of Mossad was Meir Amit. He had the same background of active service in Israel's cause, but he agreed with Ben-Gurion that Mossad should renounce direct action in favour of a more modern approach based on subterfuge and stealth. Born in Palestine, Amit had been raised on a kibbutz and then recruited into the Haganah. He fought with distinction, and was wounded, during the 1948 war, and was then given a brigade to command. Later he was transferred to Military Intelligence. He went to Washington in the late 1950's to cement the already excellent relations between Israeli Military Intelligence and the CIA. On his return to Tel Aviv in 1961, he was appointed head of Agaf Modiin – usually shortened to Aman – the espionage section of military intelligence. It was there that he learned to appreciate the immense value to the spy business of computer technology.

In the late 1950's, a brilliant Israeli scientist named Yuval Ne'eman had been assigned to Aman. Professor Ne'eman recommended that Aman should gather every single scrap of information about Israel's military enemies and feed it into computers. Even the most banal items produced from interrogating captured soldiers and security suspects, down to their sock size and the brand of tea they drank, were collated remorselessly. In that way, Ne'eman predicted, a picture of the intentions of Israel's enemies would emerge.

When he arrived at Mossad, Amit was thoroughly convinced of the value of computerized information. He borrowed the Ne'eman technique and Mossad's own files were in turn computerized. At the same time Amit equipped Mossad with the best technological gadgetry of modern espionage, such as electronic listening devices that can overhear a conversation half a mile away.

Amit found that Mossad's embattled past also had several techniques to offer. In the desperate scramble for weapons during the 1948 war, for example, Mossad had experimented with the use of sympathetic foreigners and front companies. Thus, in Britain, a Mossad agent had come across four Bristol Beaufighters, long-range nightfighters that had flown during

World War Two but had since been put up for sale as scrap. British regulations forbade the planes to take off and, anyway, rationing made petrol nearly impossible to obtain. Then a British company named Alpha-Film obtained permission to fly Beaufighters for a movie scene. As empty cameras turned the four planes took off, headed east, and disappeared. Two days later they were in action against Egypt.

In the USA, Mossad pulled off an even more remarkable coup. An arms dealer was persuaded to part with four B-17 Flying Fortresses in the belief that they were to fly for a new airline in Central America. They, too, took off and headed for Israel, one of them making an immediate contribution to the war effort by bombing King Farouk's Palace in Egypt on the way home.

These were techniques, Amit realized, which could profitably be updated to match the circumstances of the 1960's. Mossad set about recruiting friends and allies who would be able to help Israel in times of need. The hour of the square-jawed, trench-coated secret agent was past. Mossad's helpers might be drawn from the ranks of respectable businessmen with the excuse to travel and easy access to government bodies and international corporations. They could be lawyers and accountants experienced in the game of setting up companies in such improbable business centres as Liberia and Panama. They could be fast moving entrepreneurs who could offer a range of unlikely contacts, accommodation addresses and safe houses. Or they could be hucksters living on the edge of the law who would launder money to finance overseas escapades.

Amit stayed with Mossad until the 1967 war was most satisfactorily concluded, and then left to run a Tel Aviv metal firm. (The move to a high position in business is a customary route for senior officials who have served Israel well.) The Institute inherited by Zwi Zamir had been thoroughly modernised. Given that many of its top men had been with Mossad since the beginning, it was inevitable that its operations were still marked with a certain panache. But it no longer relied for success solely on its flair for brilliant improvisation. It had all the techniques of modern espionage at its disposal. The best Mossad schemes were those for which the groundwork had long been laid.

When in 1967 Mossad received its most vital assignment, the objectives were clearly stated:

- to acquire by stealth 200 tons of uranium, which was the minimum amount Israel needed to build an effective nuclear deterrent;
- to leave behind no shred of evidence that could prove Israel was the culprit.

Mossad decided to meet Dimona's needs from Belgium where a commercial supplier had the largest privately-owned stockpile of uranium in the world. That supplier was actively seeking buyers. And there was the added attraction that Belgium was a member of the European Economic Community, whose nuclear safeguards were less than rigorous. Even so, buying the uranium, getting it out of Europe, and getting it 4,000 miles home would require all Mossad's flair and audacity.

The plan Mossad arrived at had five major elements. The Institute had to:

- recruit a 'friend' who had sufficient cover to be able to buy uranium openly;
- find an excuse to get the uranium on to the high seas;
- provide the means to do so – a ship and a crew;
- give the ship a legitimate reason to take the uranium all the way from Belgium to the Mediterranean without raising the alarm;
- arrange the transfer of the uranium at sea for the last stage of the journey to Israel.

Finally there was the matter of a codename. Mossad likes to give its operations cryptic, sometimes unfathomable titles. When it set out to provide Dimona with enough nuclear fuel to last a decade, the operation was codenamed Plumbat. No modern dictionary contains such a word. It might have been intended as a clever play on words: the aim of the operation was to derive uranium; the derivative of uranium is lead; and the medieval Latin word for lead was *plumba*. On the other hand, perhaps Mossad was enjoying an obscure joke. In historical English the words 'plumb' and 'bat' were both slang expressions for deception – and deception was certainly the foundation of Mossad's scheme. Or PLUMBAT may simply have been chosen as a likely sounding trade name for barrels of chemicals, which is how the uranium was to be disguised on its journey home.

Chapter 3
A Small Town in Germany

In the secret service world, all things are possible and nothing is what it seems. It is a looking glass world, as John le Carré has said, where conventional morality has no meaning and all customary values can be overturned. Provided that exposure and public embarrassment can be avoided, the most implausible friendships can be struck, no matter how contrary they run to acknowledged international alliances.

Yet if anyone had suggested in 1945 that one of the most valuable allies of the future state of Israel would be West Germany, that would have been hard to swallow, even on the far side of the looking glass. But when Israel sought the means to build nuclear weapons, two of the men she relied on most were Germans, both of whom had fought for Hitler's Reich in the Second World War. Since the war, they had re-established themselves by dint of hard work, and, occasionally, bizarre business deals. They had formed a partnership which by chance had placed them in a pivotal position to help Israel achieve her nuclear ambitions. In such circumstances it was understandable that the horrors of the past could be forgotten.

The senior partner, at least in terms of age, was Herbert Scharf. In his eighties, and despite a recent heart attack, Scharf is a tall, handsome man with a magnificent Aryan head set squarely on broad shoulders. In 1945 his physical strength, together with undoubted determination, were almost all he had to equip him for a new start.

For Scharf, the war ended disconcertingly. Like many other career officers in the German army, he was far from being a committed Nazi. But he had accepted Hitler as his fuehrer, and spent the war attached to the headquarters of the general staff of the Wehrmacht, doing what he saw to be his duty. Germany's defeat left him in uncomfortable limbo. The victorious Allies disbanded the German armed forces and the only career Scharf had known was abruptly ended. Scharf was more fortunate than some of his brother officers who found themselves held in Allied prison camps – but barely. Approaching the age of fifty, he was homeless and unemployed, and had neither money nor prospects. He decided to leave Germany, not as a fugitive, but as one of the dispossessed.

Scharf chose to look for a new start in the Horn of Africa, in the former Italian colony of Eritrea. In 1945 it was in chaos. The Italian dictator Benito Mussolini had used Eritrea as his base for the invasion of Ethiopia and Somaliland in 1936, and for the next five years Eritrea formed part of the new territory which Mussolini grandly named Italian East Africa. In 1941 British forces advancing from the Sudan routed the Italians and Eritrea was administered by the British army until the end of the war. Then its fate was left to the Allies' military control, composed of Russia, the United States, France and Britain, which failed to reach any decision.* The confusion and uncertainty that resulted made Eritrea the perfect place for an unemployed German officer to seek refuge and a new career.

Scharf headed for Eritrea's capital, Asmara. Its cool position, perched on the eastern edge of the Hamasen Plateau at 7,800 feet, made it a tolerable place to live, and on clear days you could see across to the Red Sea, forty miles away. Scharf began trading in costume jewellery and semi-precious stones: business was done in the restaurants and bars of Asmara's European zone. (There were also 'native' zones but Europeans hardly ever needed to visit them.) There were plenty of other like-minded free-wheeling businessmen with whom Scharf could drink coffee and talk endlessly about the opportunities opening up. Scharf met a Frenchman named

*Eventually the United Nations was called upon to decide Eritrea's fate. It prevaricated until 1952 when Eritrea was unwillingly federated with the Ethiopia of Emperor Haile Selassie. The consequence of that decision was to be the bloody war that raged in Ethiopia through 1976, 1977 and 1978.

Marcel who told him about algin, a chemical obtained from seaweed and becoming widely used in the food, cosmetic, pharmaceutical and textile industries. Marcel also introduced him to the dyestuffs business. Scharf started selling algin that was produced by a Casablanca company named Chimagar and dyestuffs supplied by a French company, Francolor.

By 1948 these various endeavours had given Scharf the start he wanted. Business horizons in Eritrea were limited, so Scharf collected his savings and returned to Germany.

He chose to settle in Wiesbaden, an unhurried residential town on the north bank of the Rhine, twenty miles west of Frankfurt. Under the Allies' shareout of occupied Germany, Wiesbaden fell under US military control. Scharf soon founded his own company: 'Herbert G. Scharf, Chemisch-Technische Zentrale', which occupied the ground floor of a modest building at 30, Hildastrasse. Scharf, who confined himself to selling other companies' products, became a sales agent for the General Dyestuff Company and the General Aniline & Film Corporation of the United States. His Asmara contacts remained useful, too: he continued to sell dyestuffs for Francolor and algins for Chimagar.

The new business did well. But Scharf was left feeling unsatisfied. He wanted to be a manufacturer in his own right, but he recognized that he lacked the technical know-how. In 1950 he met a man who possessed some of the necessary qualifications, and, more important, an abundance of ambition. His name was Herbert Schulzen.

Boyishly good-looking, with a shock of dark hair hanging low across his forehead, Schulzen was a textile engineer who worked at that time for a chemical company named Benckiser in Ludwigshaven, a town on the west bank of the Rhine some fifty miles south of Wiesbaden. By chance, Scharf also sometimes worked for Benckiser, acting as a freelance sales agent for some of the company's products, and it was at Benckiser's offices that the two happened to meet. Even though Scharf was nearly thirty years older than Schulzen, the two soon became friends. Scharf was attracted by Schulzen's enthusiasm and expertise. Perhaps regarding the younger man as his protégé, Scharf suggested they join forces in the hope of one day starting their own manufacturing company. Schulzen left Benckiser; the two men started selling together; and the partnership that was eventually to prove so vital in the Plumbat story was formed.

Scharf and Schulzen were an apt team. Scharf was an earnest, dependable man; Schulzen provided the counterpoint with his quick-witted and extrovert nature. In contrast with Scharf's wartime service at general staff headquarters, Schulzen had spent his war in action with the Luftwaffe. He had gone into battle at the age of twenty, flying against Russian tanks on the eastern front. Somewhat surprised at coming through that experience unscathed, he was transferred from his fighter squadron to the Army Cooperation Flight Training Centre in Pomerania. There he became one of the earliest jet pilots, learning to fly Messerschmitt 262s. Hitler had intended to use the 262 in high-speed bombing raids on England but the plan was thwarted by a lack of fuel. It was just as well for Schulzen because the 262 was never technically an entirely happy aircraft. 'Thank God we never went into battle with them,' Schulzen said in retrospect. For him the end of the war was slightly less catastrophic than it had been for Scharf. Schulzen's father worked in the textile industry and he was able to find his son an opening.

From the start, Scharf's partnership with Schulzen flourished. Continuing to sell on behalf of other companies, they steadily built up their business. Schulzen's experience of the textile trade helped him to establish some useful – and significant – agency agreements. (One was with an Italian company which dealt in dyes for textiles. Named SAICA, that company was also destined to play a curious but important role in the Plumbat affair.)

The two men's ambition to become manufacturers in their own right grew stronger and in the early 1950's the time seemed ripe. West Germany was reorganizing from the chaos of defeat. The country was beginning to get back on its feet economically, markets were expanding for every product imaginable, and the door was wide open for ideas and enterprise. It is said in Wiesbaden that the man who came up with the best idea for Scharf and Schulzen was an ambitious young entrepreneur named Ingo Trepel, today a major West German industrialist. At that time he too was just starting in business, as a manufacturer of simple hydraulic lifting equipment.

One evening in 1951, Trepel and Schulzen were having a drink together in Wiesbaden's Parisiana Bar, now a down-at-the-heels strip club, but then an elegant revue bar. Schulzen asked Trepel if he could think of anything he and Scharf might manufacture – always bearing in mind their very limited capital. Trepel did not claim to have any original ideas. But he did point out that the workmen he employed used one barrel of industrial soap each week to clean the grease off their hands. Perhaps an opening lay there? Schulzen and Scharf embraced the idea enthusiastically. They found premises on the ground floor of 54 Am Schlosspark, Wiesbaden, in the same building, off a cobblestone courtyard, where Trepel had his business. On April 10, 1952, Scharf and Schulzen registered their new company. In memory of the fresh start Scharf had won in Eritrea, it was named Asmara.

Wiesbaden may at first have seemed an unlikely place to make and market industrial soap. One of the oldest spas in Germany, with twenty-seven hot springs, its tourist attractions also included well-ordered promenades and parks, a casino, and locally-made champagne. But in the 1950's a considerable industrial zone, with iron foundries, metal and concrete works, chemical and textile plants, was growing on Wiesbaden's outskirts. There Asmara found plenty of customers.

It began, by all accounts, in hand-to-mouth fashion. Asmara mixed one load of soap and sold it, and with the proceeds bought the ingredients for the next load. But in time Asmara began adding to its range of products. First there was *Handwaschpaste*, a concentrated soap in paste form, sold to housewives with the claim that it was 'neutral, mild, kind to the skin and hygienic'. Then came *Waschkonzentrat*, a truly remarkable concoction. According to Asmara's brochure, this 'combination of modern synthetics' could clean anything from the naked lady who appeared on the brochure's cover to aluminium, men's suits, overalls, asphalt, cars, bathtubs, printing presses, gears, hair, dogs, linoleum, furniture, carpets, horses and much more besides.

The advantage of *Waschkonzentrat* from Asmara's point of view was that like the company's other products, it was easy and cheap to manufacture. Asmara simply bought the ingredients, mixed them together and packaged the result.

But Scharf and Schulzen were now nurturing ambitions that went far beyond this humble enterprise. Ringing Wiesbaden at that time were several important United States military bases, including the headquarters of the United States Air Force in Europe. These bases generally got all their supplies, except for some fresh foods, from the US. (It was a policy which helped the US balance of payments, and also avoided over-inflating local economies.) But military procurement departments were permitted to buy locally if a manufacturer could offer a product at a cheaper price than an American supplier. That was the vast and lucrative market that Scharf and Schulzen now cast a longing look at: even a small order, in US terms, would give a gigantic boost to Asmara's fortunes.

It was not an easy market to break into. But in 1961 Scharf and Schulzen hired two salesmen with persuasive reputations. One was a suave, personable man named Wilhelm Bargon, said to be very effective at breaking down the resistance of secretaries in outer offices. The other was Herbert Lettko, a chemist by training, who had already won entree to the US bases around Wiesbaden through his sheer and dogged persistence: 'The sort of guy,' as one of his friends put it, 'who if you say no to him and lock the door, he comes in through the back. If you throw him out the back, he'll come in through the window."

They were a formidable pair, and were soon selling Asmara's products to US bases. For good measure they penetrated the procurement office of the West German army, and before long Asmara was firmly in the ring of companies that thrive on military patronage.

In all honesty, it was not a very sophisticated business. There is the fairly typical story of one German dealer who won a highly profitable contract to supply the US Air Force with petroleum ether, commonly used as a cleaning solvent. The Air Force insisted, in its contract, that the ether be supplied in plastic containers which it had already bought from another supplier. The dealer pointed out that these very expensive containers would almost certainly be eaten away by the petroleum ether, which would become contaminated in the process. His advice was ignored. He carried out the contract to the letter, buying a large quantity of the solvent, bottling it in the containers which the Air Force supplied, and adding his own fancy label. The containers and the label cost five times as much as the ether, which soon became contaminated exactly as predicted and had to be thrown away.

Asmara's own military deals were mostly of a similar nature. Once it too won a large rebottling contract, agreeing to supply chloramine, a mild disinfectant, in small containers. Asmara had to rent a barn to carry out the operation. Again, the cost of the containers and the label far exceeded the cost of the contents.

Asmara's military sales had not yet managed to transform it into the substantial German company Scharf and Schulzen so desired. But in 1962 Herbert Scharf returned from a business trip to Paris, convinced that he had stumbled across the answer. While he was waiting at Orly Airport, a plane had arrived after somehow becoming contaminated by radioactivity. Scharf had watched the plane being washed down with generous amounts of chemical foam.

In Wiesbaden, Scharf assembled Schulzen and the two salesmen Lettko and Bargon and asked: 'What is decontamination?' No one present knew, but that did not deter Scharf, who announced, 'We are into decontamination – that's our business now.' There remained the small problem of finding out exactly how to make decontaminants, but that was solved when Asmara acquired several Russian textbooks on chemical and atomic warfare from East Germany. There in black and white were chapters on decontamination, including the formulas for making creams and ointments to offer some protection against nerve gas, blister gas, and radioactive fallout.

With these formulas as its starting point, Asmara went into the business of manufacturing decontaminants. Within eighteen months, it had captured its first big order from the German Defence Ministry worth about half-a-million dollars.

Schulzen in particular became something of an expert on the subject. He distinguished himself by writing an article on 'Chemical Sabotage Poisons' for a civil defence magazine and was able to allude to his impressive contacts by thanking, at the end of his article, Herr Ludwig Scheichl of the German Defence Ministry for 'the numerous suggestions and the support rendered.' Schulzen also told his friends that he had become a consultant on chemical warfare to NATO. Meanwhile, Asmara was developing a range of personal decontamination sets which it sold to the armies of several countries.

Thus, by the mid-1960's Asmara Chemie had won itself a tidy niche in the defence market, and had built up a useful range of industrial and military contacts. But as a company it was still small enough to remain flexible and versatile. It was at this point that Asmara, and the ambitious Herbert Schulzen, came to the attention of Israel.

When Israel was born in 1948 it would have been unthinkable to most of its citizens that the new state should cultivate the friendship of men like Herbert Schulzen. He was German, he had served the Third Reich and he stood on the wrong side of the widest river of blood in history. The public reaction to the first tentative encounters between Germany and Israel demonstrated the problems their leaders would have to surmount. In 1952 the mere suggestion that German and Jew might sit at the same table to negotiate compensation for the six million victims of the holocaust brought battle scenes to the streets of Jerusalem. For hours crowds stoned the Knesset while inside the chamber members made the pretence of continuing the debate with tears pouring down their cheeks from the tear gas that drifted in through the broken windows.

Eventually Israel agreed to accept three billion Deutsch marks in reparation. Israel's dire economic needs had made some such accord inevitable. Even so most Israelis could not forget the past, would not forgive it. After he had played the music of Richard Strauss in public, an Israeli violinist was stabbed in the street. When the first prominent West Germans visited Israel there were demonstrations and further emotional debates in the Knesset. And twice Israel's Prime Minister David Ben-Gurion was forced to resign, bringing down his government with him, in consequence of his belief that Israel must, without forgetting the past, build some kind of future with Germany.

Yet by the early 1960's Herbert Schulzen, and Germans like him, were not merely acceptable allies; they were sought after.

Part of the explanation for this remarkable change in Israeli attitudes lay, of course, in the simple passage of time which healed the wounds if not the scars. Increasing numbers of West Germans were allowed to visit Israel: first, there were carefully-screened tourists; next young Germans who went with unassailable sincerity to expiate their parents' crimes by working in *kibbutzim*. And then came the agony of the long, drawn-out trial of Adolf Eichmann, with its testimony, provided by 100 witnesses, to the utter depravity of Hitler's Nazis, which acted as a kind of catharsis. The memories remained, but for most the torment was gone.

In this milder climate, the backgrounds of men like Schulzen became less significant barriers to friendship. And if this natural healing process needed any impetus, it came when Israel saw yet again the fragile nature of the bonds of official alliances.

In 1959, with Mossad's help, Israel began to explore the possibility of obtaining military aid from West Germany. The moves were conducted in conditions of utmost secrecy. This time it was West Germany who had most to lose: one false step could wreck her relations with the Arab nations of the Middle East. Late that year Israel's Defence Minister Shimon Peres paid a clandestine visit to his West German counterpart Franz Josef Strauss. Peres flew to Paris where he rented the smallest car he could find and drove through ice and fog to Strauss' home south of Nuremburg. There Peres told Strauss that Israel needed Germany's help. The United States was giving Israel money; France was selling Israel arms; 'Germany,' Peres said, 'would be taking a far-sighted step in building bridges to the past if she would help with arms without requiring either money or anything else in exchange.' Strauss, who said he was concerned about a possible 'back door' Soviet attack on Europe via the Middle East, was sympathetic. In February 1960 the deal was sealed in an equally top-secret meeting between West Germany's Chancellor Konrad Adenauer and Israel's Prime Minister David Ben-Gurion in New York.

Soon Germany was supplying Israel with small arms, ammunition, helicopters, training equipment, and spare parts. By the end of 1963 Germany had agreed to supply tanks and the two countries were discussing some small submarines Israel said she wanted. But in March 1964 came disaster: The Arabs found out. Such had been the secrecy surrounding the arms transactions that several West German ambassadors were trapped into denying the first reports of the deal, honestly believing them to be false. They soon had to retract. The consequences for Germany's Middle East policy were catastrophic, as many of the Arab states broke off diplomatic relations with Bonn and threatened to recognize East Germany into the bargain.

West Germany told Israel the arms supplies must end. To compensate, Germany gave Israel money and formal diplomatic recognition. Shimon Peres was moved to point out, in some bitterness, that neither cash nor the presence of a West German ambassador in Israel would offer much protection against Arab bombs and missiles.

The lesson of that disappointment was, once again, crystal clear: to survive, Israel must continue to cultivate friends who would not, like governments, be easily swayed by the shifts of international politics.

Making contacts with individuals, groups, organizations and companies throughout the world has become a national cause in which every Israeli and every Jew can participate. Occasionally, Mossad itself plays an active role. For example, when West Germany reneged on its military aid agreement and refused to supply some secondhand tanks it had promised, Mossad found an Italian company willing to buy them 'for overhaul.' The tanks made the briefest of visits to the Italian factory, and were then shipped secretly to Israel.

More often Mossad simply monitors the progress of likely-looking relationships with foreigners, storing information in its computerized files in Tel Aviv until Israel requires a suitable friend to fulfil one of its special needs. The motivation of these friends is varied: ideology, either political or religious; sympathy or guilt; or less prosaic forces such as self-interest or money.

There are many ironies in the Plumbat story. One of the greatest is the probability that Herbert Schulzen's relationship with Israel came about as a result of injuries he sustained while fighting for the Third Reich.

Schulzen's wartime career as a Luftwaffe pilot had met an abrupt and painful end in a field in Denmark in 1945. After taking off from an airfield near Flensburg on the German-Danish border he lost a brief encounter with a Royal Canadian Air Force Mosquito. In the consequent crash landing he suffered a serious head injury and his wound troubled him increasingly over the years. Finally, in 1964, he was advised to undergo major surgery, and it was while he was convalescing that the first approaches were made to him on Israel's behalf.

Who spotted Schulzen is uncertain, but there is a remarkable coincidence which suggests that it may have been Dan Aerbel, the Mossad agent who was arrested after the Lillehammer murder in July 1973. The documentary evidence proves only that Aerbel became involved in Operation Plumbat towards the end. But it was in 1964 that Aerbel was recruited by Mossad – and he spent a great deal of that year in Wiesbaden. Using the cover of a furniture salesman, Aerbel visited the US military bases around Wiesbaden with which Schulzen's Asmara was doing an increasing amount of business. Part of Aerbel's job was, undoubtedly, finding likely 'friends' for Israel and Schulzen would have seemed a promising prospect: he had good contacts in the industrial and military fields; he was highly ambitious; and he was a partner in a company that was modest enough in size to be extremely flexible. There was the added factor that Schulzen would probably welcome, as part of his convalescence, a relaxing stay in a Mediterranean climate.

In any event, Schulzen was invited to Israel, ostensibly by a Tel Aviv company whose activities included manufacturing furniture. His visit was convivial. He made many friends, who jokingly nicknamed him 'the Nazi pilot,' and he was shown around the prestigious Weizmann Research Institute at Rehovot. When he returned to Wiesbaden, fully recovered from his operation, he brought back warm memories and a picture book of Israel, fondly inscribed by one of his new acquaintances.

It was not long before Schulzen's visit brought Asmara tangible rewards in the form of orders from Israeli companies. Sometimes they came direct; sometimes they were channeled through other German companies, other friends of Israel, to which Asmara paid a commission. It was totally straightforward commerce, with Asmara providing such items as large quantities of Tuftophil K, a special chemical softener used in textile manufacture.

But Asmara's name had now gone down in Mossad's files to await the day when Israel might have more urgent needs than textile softeners to fulfil. That day was not long in coming.

The June war of 1967 left Israel facing embargoes, some official, some not, on many of the vital supplies she depended upon. She now turned to some of her private friends who might be willing to help bridge the gaps.

As it happened, June 1967 also saw some significant changes at Asmara. Herbert Scharf, approaching seventy, decided formally to surrender some of the responsibility for running the company, and thus give Schulzen more control. On 19 June the company's three directors – Scharf, his wife, and Schulzen – awarded Schulzen 'sole powers of representation,' which, in plain language, gave him the power to make deals on Asmara's behalf. And it was shortly after this alteration that the company's business with Israel took on a distinctly military hue.

First Asmara supplied 300 of its personal decontamination sets to the Israeli army. Next it attempted to provide advanced aerial photography equipment. Ernst Richartz, a writer from the Wiesbaden area who sometimes dabbled in buying and selling, was asked to go to Schulzen's home in Hettenhain. Only a village really, Hettenhain is tucked away amidst pine forested hills northwest of Wiesbaden. Schulzen's well-appointed home, built on the side of one of those hills, is surrounded by a tall, solid wooden fence to keep out prying eyes. Richartz was admitted and asked to take a seat by the small swimming pool that occupied nearly the whole front garden. Schulzen asked him whether he could help Asmara acquire some infra-red aerial cameras, a subject Richartz knew something about. The deal came to nothing in the end but Richartz was left in no doubt that Asmara was deeply entwined with Israel.

And then came the biggest, the most spectacular deal of all. Asmara agreed to act as the 'front company' in the acquisition of 200 tons of uranium. Schulzen was jubilant. It was by far the largest order Asmara had been involved in, and he could not resist boasting about it. Early in 1968 he bumped into Herbert Lettko, the persistent salesman, in Wiesbaden. Lettko had left Asmara by then to set up his own company but the two men had remained friendly. Schulzen told Lettko that he had just pulled off a major deal with Israel. He told Lettko that the deal involved *urea.* Lettko thought that Schulzen was referring to a compound used as a fertilizer. It was only later that he realized that this was a somewhat laborious joke which depended on the similarity between *urea* and *uran* – the German word for uranium.

Chapter 4
A Valuable Commodity

Olen is a pretty red-bricked village twenty-five miles east of Antwerp that drums up a respectable tourist trade on the strength of a somewhat implausible Flemish legend. It is said that in the sixteenth century King Charles I of Flanders and Saxony stopped to refresh himself with a tankard of beer at one of Olen's taverns. The traditions of local hospitality insisted that the landlord should hold the pewter tankard by its handle when passing it to his guest; the king's dignity required him to receive the tankard by grasping the same handle. A blacksmith was hastily summoned to equip the tankard with a second handle but the landlord now insisted on holding both. The impasse was ended only when the blacksmith added a third handle for the king. Today every tavern in Olen displays what is claimed to be the original three-handled pot. Replicas are for sale, of course.

The more serious side of life at Olen is situated some way from its tourist attractions. On the bank of a wide canal five miles from the village centre are the smoke-belching chimneys and gigantic black slag heap of a metal refinery. It is owned by the Belgian company Metallurgie Hoboken-Overpelt, and it is the only place in Belgium licensed to store radioactive materials. It was there, in a silo, that Israel's uranium was waiting.

The order for the uranium arrived at the Brussels headquarters of the Belgian company Societe Generale des Minerais in March 1968. SGM, a sister company of Metallurgie Hoboken-Overpelt, is part of the giant conglomerate Societe Generale de Belgique, with interests ranging from banking to shipping, paper-making to mining. As befits its status at the centre of the Belgian economy, SGM occupies a prestigious and somewhat old-fashioned office building in the Rue du Marais in central Brussels. It is a solid, stone building with calm, wide corridors, heavy wooden doors, polished mahogany furniture, and cut-glass ashtrays. The uranium order was directed to the large and comfortable office of Denis Dewez, deputy head of SGM's uranium division.

Dewez knew it was an order that SGM could easily fill. Another of SGM's sister companies was Union Miniere, a mining corporation which for a long time was the power behind Belgium's rule over her African colony, the Belgian Congo. Belgium had pulled out in 1960, bequeathing a bloody civil war as different factions vied for political control and the Congo's rich mineral resources. By staying on until 1965 Union Miniere managed to ship home large quantities of those resources, including copper and a substantial amount of uranium oxide. It was that uranium which was stored in the silo near the village of Olen. SGM was having trouble finding buyers to help diminish the small mountain. In 1968 there was little military demand because of the nuclear test-ban treaty passed five years earlier, and little civil demand because peaceful nuclear programmes were still in their infancy. Even so, Dewez regarded the order for 200 tons with considerable caution.

His wariness stemmed from the fact that the would-be buyer was Asmara Chemie of Wiesbaden. Nobody at SGM had ever heard of Asmara Chemie. And since the transaction would undoubtedly involve several million dollars, Dewez was naturally curious to know whether this unknown company was in a position to pay. When Dewez initiated a polite inquiry to that effect, Asmara's reply was reassuring. Not only was it able to pay, the necessary funds were already lodged with a bank in Zurich, waiting to be transferred at the appropriate moment to SGM's account. The bank itself confirmed that this was true. But no one in SGM had heard of the bank, either. SGM had not achieved its position as one of the biggest non-ferrous metal dealers in the world by casting aside prudence, so it caused circumspect inquiries to be made into the bank's financial standing. When those inquiries resulted in satisfactory references, SGM said it would be happy to deal with Asmara. The reasons Asmara had given Dewez for wanting to buy the uranium seemed plausible, and SGM saw no need to inquire any further into the standing, financial or otherwise, of Asmara itself. In the circumstances, that was just as well.

The story Herbert Schulzen had told Dewez went like this. Asmara was about to engage in the mass production of petrochemicals. In that endeavour it intended to use the uranium as a catalyst to begin or accelerate the chemical reactions which this type of manufacturing calls for. This in itself seemed reasonable to Dewez because in those days of comparative plenty, uranium was sometimes used in this way.

Before it can be used as a catalyst, however, uranium needs to be treated, a fairly complex process which Asmara explained it was not capable of carrying out. Asmara told Dewez it had arranged for its uranium to be treated by a chemical company called Chimagar of Casablanca in Morocco. Asmara proposed to ship the uranium to Morocco, and then after treatment ship it back to Wiesbaden.

Chimagar was the same company which had helped Herbert Scharf make his new start in Eritrea. And in the design of the scheme the Moroccan connection was an ingenious touch. Had this part of the plan come off, and had there been a hue and cry following 'the loss' of the uranium, then suspicion would have fallen not on Israel but on the Arab world. Unfortunately, the little joke misfired. As Dewez now innocently pointed out, Asmara seemed to have overlooked one inconvenient fact of European life.

Germany was part of the European Economic Community which then consisted of six nations: as well as Germany, there were France, Belgium, Italy, the Netherlands, and Luxembourg.* Movements of uranium into and within those countries were controlled by an EEC agency called Euratom. As nuclear watchdogs went, it was a pretty pathetic creature, with no bite and very little bark. But exporting uranium from the community even on a temporary basis, which is what Asmara said it had in mind, was a more

*In 1973 the EEC was enlarged to include Great Britain, Denmark, and the Republic of Ireland.

controversial business. Euratom on its own did not have the authority to sanction exports and would have to seek permission from the EEC's executive, the European Commission. As this was an essentially political body incorporating many different shades of opinion, the Commission tended to be exceedingly cautious. Dewez warned Asmara that since Morocco was not a member of the EEC, obtaining permission to export the uranium to Casablanca would be neither quick nor easy.

Asmara appeared to accept the setback with stoicism. It told Dewez that it would look around for another company within the European community capable of treating the uranium. Meanwhile, Asmara said, there was no reason why negotiations over such matters as price, delivery and so on should not continue. They did, very satisfactorily, until Denis Dewez unwittingly cast a sizeable wrench in the works.

Up to this point, the late summer of 1968, the fiction that Asmara was going into the mass petrochemical business had been easy to maintain. All the negotiations had been conducted by letter or telephone, leaving SGM happily ignorant of the fact that Asmara was a very modest little company. But then Dewez called Herbert Schulzen to suggest that the time had come for some face-to-face bargaining. And since Asmara was the customer, Dewez said it was only right that he should accept the inconvenience and expense of travel. He proposed politely to Schulzen that he should visit Wiesbaden to continue the talks.

It was a decidedly nasty moment. Schulzen could hardly decline Dewez's courteous offer. But neither could he possibly let Dewez visit Asmara's premises. Once Dewez saw for himself the cobbled courtyard leading to the tiny entrance and cramped offices at 54 Am Schlosspark, he would have realised the impossibility of even finding room there for 200 tons of uranium, let alone doing anything with it.

It was a time for quick thinking, and the quick-witted, extrovert Schulzen, with his long experience as a salesman, met the crisis. He responded to Dewez's suggestion in the most gracious way possible – by inviting Dewez to his own home in the village of Hettenhain, ten miles or so from Wiesbaden. With a swimming pool and three-car garage, it would certainly give Dewez the impression that Schulzen was a man of substance. And fortunately for the Plumbat scheme, which could have foundered at that moment, Dewez was not burdened with a suspicious nature. He accepted Schulzen's invitation without further inquiry or complaint. The crisis passed.

The meeting itself went smoothly enough. There were several men present besides Schulzen whom Dewez assumed to be other representatives of Asmara, including, he supposed, at least one scientific or technical expert. He further assumed that they were all Germans, since that was the language they spoke to each other. As Dewez himself did not speak German, the actual negotiations were conducted in English. From Schulzen's point of view the ground they covered was safe: such matters as the quality of the uranium, the delivery date and the price. And when it came to the unfortunate snag presented by the Moroccan company Chimagar, Schulzen was able to announce that he had now found another company willing and able to undertake the processing of the uranium. The company was in Milan, Italy. Since Italy was a member of the EEC Schulzen supposed that official permission to send the uranium to Milan would be easily obtained. Dewez agreed.

There was just one tiny snag, Schulzen said. Italy was a long way from Belgium and he felt the easiest way to get such a cumbersome load there would be by ship; however, that would mean the uranium leaving the EEC for the length of the voyage. Would that technicality present any problems? Dewez thought not. A voyage on the high seas did not really constitute exportation. In any event, Dewez said, when he returned to Brussels SGM would draw up a contract which could be submitted to Euratom for approval. As he left Hettenhain to return to Brussels, Dewez assured Schulzen that he was fairly confident that approval would be forthcoming.

Just as SGM had felt no curiosity about Asmara's background, it now displayed no interest whatsoever in the credentials of the Italian company. That, too, was fortunate. Schulzen had chosen the company on the strength of a personal friendship with its owner, Francesco Sertorio, that went back twenty years. Sertorio, a tall and aristocratic looking Italian, had readily accepted Schulzen's proposition that he be named, in lieu of the Moroccan company, as the recipient of 200 tons of uranium. The fact that Sertorio's company had, like Asmara, never handled uranium and had no idea how to process it, did not bother either man one little bit.

The beauty of Operation Plumbat lay in the way it was designed to exploit weakness. Above all, of course, Mossad was relying on the weakness of the European safeguards that were supposed to prevent uranium from falling into the wrong hands. But there was also the more subtle way in which the designers of the plot preyed on human frailties, such as avarice. A man in trouble does not usually interrogate the bearer of any good tiding. When Schulzen had arrived in Milan, bearing the uranium proposition, Sertorio, or at least his company, was in trouble.

Societa Anonima Italiana Colori e Affini (SAICA) had been founded in May 1934 to trade in ships' varnishes. When the varnish business went into recession, the company had switched to dealing in dyes for the textile industry. It had barely had time to become established before the commercial strains of World War II all but killed it.

After the war SAICA had limped along until Sertorio invented an ingenious method of mixing dyes which gave printed textiles a brighter hue. He took out patents on his idea and was able to export it to several countries, including West Germany, earning the company substantial royalties. On the strength of that SAICA had been able to sell some of its shares to the Italian public, and in recognition of his achievement Sertorio had been appointed managing director. But SAICA suffered from the handicap that it did not actually manufacture anything; it was simply a trading company. And by the mid-1960s, when Sertorio's original patents began expiring, trade was none too good.

In 1965 SAICA's profits were around £700* before tax

*All sums of money are stated in pounds sterling. The conversions from other currencies are based on the exchange rates prevailing at the time.

which, as Sertorio said in the company's annual report, 'must be considered really modest.' No dividends were paid that year and 1966 was no better. The profits did creep up to £900 or so but there were still 'tax problems' and no dividend.

It was much the same story in 1967, and again in 1968 when matters seemed to reach a catastrophic climax.

Sertorio's main partners in SAICA were two brothers named Giancarlo and Franco Campari. On 19 March 1968, Giancarlo was killed in a car crash and soon afterwards Franco had a blazing argument with Sertorio and quit.

It was around this depressing time that Schulzen approached Sertorio with the uranium proposition. Asmara had represented SAICA's interests in West Germany for some time and Schulzen and Sertorio had become close friends, perhaps drawn together because both were considerable extroverts. Sertorio's son Stefano remembers that 'Schulzen used to make us laugh. He was very un-German.'

In view of SAICA's precarious condition, Schulzen's proposition was obviously enticing, but there were a couple of problems standing in the way. For one thing, SAICA did not have a plant in which to process any uranium, only modest office premises in Milan. Fortunately Sertorio was also a director of a company called Chemitalia, a more substantial concern, which was in the business of manufacturing dyes and colours, and which did have a plant, of sorts, in Liainate, about ten miles from Milan. Perhaps SAICA could borrow the plant?

The second problem concerned SAICA's constitution, which under Italian law specified what sort of business the company could engage in. SAICA's constitution at that time allowed it merely to 'trade in paints and related substances,' a definition that could not easily be stretched to include 200 tons of uranium oxide. The best solution was for SAICA to adopt a far more liberal constitution and that is what it did on 26 August 1968: Henceforth SAICA could engage in 'the intimate mixing of pigments and other chemical products or primary materials of any type and for any branch of industry,' which just about covered everything.

All SAICA now had to do was learn the art of turning uranium into a petrochemical catalyst. Sertorio readily accepted Schulzen's blithe assertion that SAICA would be able to carry out the treatment process by following a set of written instructions. Schulzen said he would send the instructions with the uranium.

Chapter 5
The Mysterious Mr Yarisal

As Denis Dewez returned to Brussels to process Asmara's order, and SAICA dutifully liberalized its constitution in Milan, another act in the carefully orchestrated drama was taking place 190 miles away in Zurich, the financial capital of Switzerland. On 19 August 1968, a Swiss lawyer named Gerd Lanz was hired to set up a shipping corporation to be based in the tiny West African state of Liberia.

Lanz is a German-speaking Swiss, elegant in dress and manner. He shares his modern offices in a busy and prosperous Zurich shopping street with his son, who takes on most of the day-to-day running of the practice. Lanz has achieved this comfortable and satisfying position by hard work and application, and is not about to jeopardize it by any random violation of the ethics and traditions of his profession. Foremost of those – it is also a custom of other branches of Swiss business life, above all, banking – is secrecy. He says that he sees no reason to reveal who he was acting for when he set up the Liberian shipping corporation that features in the Plumbat story. Indeed, he goes further: for him to do so would be to break the law, which in Switzerland is most strict on this point. Lanz is a courteous man; he is also firm.

Whoever hired Lanz presumably took great comfort in the Swiss code of secrecy. There was another reason why Lanz was well-placed for the job: strange though it seems, Zurich is a centre for the business of forming Liberian corporations. Literally anyone can set up a Liberian company but Liberia insists on just one small formality. The request must be made through a lawyer – any lawyer, anywhere, so long as his or her name appears in a legal directory. Lanz clearly fulfilled that requirement. And being in Zurich, he was ideally placed to make a speedy and direct approach to the European headquarters, on Zurich's Pelikanstrasse, of Liberian Corporation Services, Inc, which exists to assist those wishing to take advantage of the country's business facilities. (It has an American office, too, on Park Avenue, New York.)

The task entrusted to Lanz was very simple to perform because Liberia's attitude to business is, to put it mildly, liberal. She is, for example, the most prolific distributor of 'flags of convenience' under which ships can be operated with the absolute minimum of interference.* In 1968 the process of placing a ship on the Liberian register was absurdly easy. All that was required was forty-eight hours notice and payment of a fifty pence fee for each ton the ship in question weighed. The ship could then hoist Liberia's red, white and blue ensign, with a single star, and sail the world in happy contravention of the mass of regulations which most other nations impose. Registering a Liberian shipping corporation was even easier. The process took just twenty-four hours and cost about £250. It was not even necessary for the shipping corporation to own a ship. Under Liberian law it could do anything from investing in real estate to exploiting inventions – a benevolent attitude towards commerce which brought Liberia a brisk trade in instant companies.

Thus when Lanz presented his credentials to the Pelikanstrasse office, no questions were necessary about the identity or background of his clients. Communications with Liberia's capital, Monrovia, have naturally been made as smooth and as fast as possible, and in only twenty-four hours a new corporation was established. Its name was the Biscayne Traders Shipping Corporation.

For the first twenty-four hours of its existence, Biscayne Traders had three directors. Their names were P. Satia, E. K.

*The credit for Liberia's prominence in this field belongs in large measure to the Greek shipping tycoon, the late Aristotle Onassis. In 1947 he was approached by Edward Stettinius, a former US secretary of state, who had set up the Liberian Trust Company to encourage US investment in what had originally been a colony for freed American slaves. Onassis proposed that Liberia should issue flags of convenience. When the idea was taken up Onassis took full advantage by placing a good proportion of his fleet of oil tankers under the Liberian flag.

Nugba and J. D. Boyd, and the corporation's address was listed as 80 Broad Street, Monrovia. That address also happens to be the head office of the Liberian Trust Company, which represents most Liberian corporations. Messrs Satia, Nugba and Boyd were three of the Trust Company's employees who made their living by acting as founding directors. Twenty-four hours later, as was their custom, they resigned from Biscayne Traders en masse.

That same day, 21 August 1968, an affidavit was sworn before the notary public of Montserrado County, Liberia, testifying that the president of Biscayne Traders and the man now empowered to act on behalf of the corporation in all matters was Burham M. Yarisal.

Of all the characters caught up in Operation Plumbat none is more elusive, more inscrutable, than Burham Yarisal. To begin with, Yarisal lives on the move: when we set out to find him, we were told that he might be in Switzerland, or Italy, or Turkey, or the Far East, or Asia, or perhaps the United States. Even people who have done business with Yarisal for a long time do not know how to contact him directly. They have telephone numbers for him in Geneva and Genoa but when they call he is always, it seems, away. They can leave messages for Yarisal but there is never any guarantee that he will call back.

The public record offers just one personal fact about Yarisal: taken from court archives in Milan, where he set up one of his companies, it reveals that he was born on 28 August 1919, in Tekirdag, a Turkish port about fifty miles west of Istanbul. For the rest it is necessary to compose a picture from the jigsaw fragments Yarisal has occasionally left behind, and put that with impressions gleaned from the few people who are willing to talk about him.

Yarisal is described by a seaman, who once met him on the dockside of the French port of Rouen, as an impressive man: short, heavyset, black hair turning grey, well dressed, a non-smoker and nondrinker. He also won respect for not indulging in the common habit most shipowners have of telling dirty stories. Another man, a shipbroker, speaks of Yarisal's fondness for animals, and his acts of personal kindness towards his business acquaintances. He is said to speak a number of languages well, including English, French, Italian and – naturally – Turkish. He has been married, but lives apart from his wife, who stays in a Geneva apartment, together with one of two daughters, who was once seriously hurt in a road accident.

In contrast there are other characteristics which have apparently made Yarisal a difficult person to get to know. He may show his acquaintances courtesy and respect, but he also has a habit of disappearing, and surfacing again only when he requires something from them. He is also given to hiring crews and laying them off at very short notice – a practice disconcerting for most professional seamen who, despite their reputation, prefer a degree of security.

This picture of a man anxious not to stay in the same place too long, determined also to leave as few clear traces as possible, is confirmed by a further search of business and legal records.

For example, the court archives of Milan – those that reveal his date and place of birth – also show that on 24 February 1970, Burham Yarisal, 'a shipowner', formed the Falken Shipping Corporation in Genoa. According to these records, Yarisal did not actually reside in Genoa or Italy at that time, but in Geneva, Switzerland – to be precise, at 10, Avenue des Amazones. On the other hand, the Genoa telephone book for 1970 lists a Burham Yarisal living at 10, Viale Pio VII – which is interesting, for the records in Milan show that to be the address of one of Falken's shareholders, a woman named Giuseppina Milena Guerra whose maiden name was Crocco. There is no trace of Giuseppina Guerra in Genoa today, but there *is* a Giuseppina Crocco. Alas, she says she has never heard of the Falken Shipping Corporation or Burham Yarisal.

Across the Swiss border in Geneva, there is also no trace of Yarisal at 10, Avenue des Amazones, a modern apartment building. But just around the corner in this chic residential suberb of Geneva, Yarisal's wife Rinka, and one of their daughters are to be found. Mrs Yarisal says that the comfortable apartment is her home; her husband does not live there. She says that he telephones her from time to time, and she offers to pass on a message. But then Yarisal does not return calls – at least not those from journalists.

Back in Genoa, the Falken Shipping Corporation does exist. It occupies a tiny office on the eighth floor of 10, Via dei Archi: the office contains a desk, a telephone, a cupboard, a telex machine, and a map of the world. It is managed by Yarisal's eldest daughter, who is married to an Italian, and refuses to discuss her father's affairs. Sometimes, however, there is an occasional helper in the office, who will give his age, sixty-eight, but not his name, and who is willing to say that Falken Shipping owns one vessel, a cargo ship, called the *Flower Bay*. Well, not quite.

Lloyd's Register of Shipping shows that the *Flower Bay* is actually owned by the Doriman Shipping Corporation of Liberia. Who owns Doriman is not a matter of public record, but its address is 'care of' the Falken Shipping Corporation, 10 Via dei Archi, Genoa. Whether or not Yarisal owns Doriman, he is certainly president of another Liberian company, the Luby Shipping and Trading Corporation. It now appears to be dormant, but in the 1960's Luby Shipping owned at least two ships, the *Linora* and the *Almaflora*. The Luby Corporation was established on 19 June 1958 . It was founded with three directors provided by the Liberian Trust Company in Monrovia, who resigned en masse the next day in favour of Burham Yarisal. It was the same method used to found Biscayne Traders ten years later.

Any attempt to piece together a coherent picture of Yarisal's activities from public records founders on these contradictions; whether or not they have been created deliberately is hard to decide. There is, however, one group of people who are not surprised by such contradictions, and, what is more, claim to be knowledgeable about Yarisal. They are the shipping brokers of Genoa.

Genoa is Italy's largest, and oldest, port. Its streets are narrow and crammed with high, ancient buildings with echoing hallways and creaking elevators. Here in jammed profusion are the offices of stevedore companies, freight handlers, shipowners, ship agents, insurers, importers, exporters, and – most of all – brokers. Brokers live by finding cargoes for ships and ships for cargoes. The essence of their game is speed and they do almost all their business by telephone. (It is not unusual for a broker to run up telephone charges of seven or eight hundred dollars a day.) Most of their deals are done by word of mouth; the ships would have sailed and the cargoes rotted long before a written contract came through. Trust is obviously vital in the brokers' trade, and by the same token so is gossip. It would be foolishness to trust someone without knowing anything about him.

Such gossip is known as 'broker talk.' Brokers claim that ninety per cent of it is accurate. 'It has to be,' said one, 'otherwise we'd all be out of business.'

This is what 'broker talk' has to say about Burham Yarisal. He got his start in business after World War II in Ethiopia, then in a state of considerable disarray. (In that respect it closely resembled its neighbour Eritrea, where – by another nice coincidence – Herbert Scharf of Asmara went to pick up the pieces of his life.) It had also become a dumping ground for the debris of war: mountains of weapons and ammunition which the Allies no longer required and the defeated armies of the Axis were not allowed to keep. There were fabulous deals to be made in war supplies and an army of scavengers invaded Ethiopia. Among them, it is said, was Yarisal.

According to 'broker talk' Yarisal based himself in the Ethiopian city of Diredawa. There he went into partnership with a fellow Turk, buying surplus equipment and selling it to the needy. That category included the Jewish settlers in Palestine, preparing, despite the British blockade, for the full-scale war with the Arabs which they knew lay ahead. Yarisal's partner died; Yarisal inherited his share of the business, which by then was substantial, and moved to Cairo. Egypt under King Farouk welcomed men like Yarisal: aged about 30, he had acquired enough capital to set himself up as an international entrepreneur, trading on opportunity. In the next three decades he was prepared to deal in almost any commodity. He won a reputation for being willing to take risks, such as trading in Vietnam and in Lebanon at the height of their respective wars, when the dangers were high, but so were the profits, especially in oil and arms. 'Broker talk' is tinged with admiration for Yarisal: 'A man like that has to keep awake twenty-four hours a day, has to stay one move ahead all the time.'

We tried very hard to check these and other stories with Yarisal. We spoke to him once by telephone, when he was in Geneva in August 1977. He told us that his background was none of our business. He said that he had never had any connection with the Biscayne Traders Shipping Corporation, or with a small cargo ship bought in September 1968 in that company's name.

The public records, and those acquaintances of Yarisal who had parts in the Plumbat affair, tell a very different story.

Burham Yarisal's first concern as president of Biscayne Traders Shipping Corporation was to find a ship. Uwe Moeller was a shipbroker in Hamburg whose firm first did business with Yarisal in the late 1950's. Moeller had bought and sold several ships on Yarisal's behalf, and also helped him make a few acquisitions of a more personal nature, such as a used car – a Mercedes 200 – and a puppy which in time became a successful show dog. Moeller is among those who found Yarisal a considerate man, given to occasional acts of kindness. Yarisal bought presents for Moeller's wife, who suffered from multiple sclerosis, and a first birthday cake for Moeller's son. Yarisal's way of life, his habit of suddenly appearing when he wanted something, then disappearing until the next time, had prevented any real friendship from developing between them. But on a business level Yarisal was a totally satisfactory client who always paid on time and without quibble.

In late August Yarisal telephoned Moeller and said that he was in the market for a cargo vessel. He was looking for a small one, of around 2,500 tons, and he was willing to pay up to £175,000.

Moeller set about the task of finding a ship for Yarisal's corporation with energy. Even so, it was three weeks before he found a candidate that fitted the specifications. Built in Rendsburg, West Germany, in 1955, she was a 258-foot cargo boat that weighed 2,620 metric tons and had a 1,650 horse-power diesel engine capable of propelling her through the water at a respectable 12½ knots. Her name was the *Scheersberg*.

As luck would have it, the owners of the *Scheersberg* were on Moeller's doorstep in Hamburg. They were August Bolten, a sizeable West German shipping company; Moeller, with Yarisal's approval, began negotiations immediately. Bolten had been trying to get rid of the ship for some time, Yarisal was eager to buy her, and the two parties had little difficulty in agreeing on a price – one and a half million Deutsche marks, or just under £160,000. All that remained was to take the customary precaution of ensuring that the *Scheersberg* was sound. As she was in Rotterdam she was put into dry dock there for a survey that proved to be satisfactory. On 27 September, on Yarisal's instructions, a Hamburg bank handed over the purchase price to Bolten. A little more than five weeks after its incorporation, Biscayne Traders had a ship.

Quick decisions and speedy deals are an integral part of the shipping business. Even so Bolten was impressed at just how smooth the sale of the *Scheersberg* had been; Yarisal must be a good man to work with. Juergen-Adolf Binder, one of Bolten's junior partners, decided that his firm should try to develop its relationship with this new client, and so he offered to find cargoes for the *Scheersberg*, in return for the usual commission. The answer he got was very surprising: a categoric 'no.' Yarisal said he had no need of cargoes.

In light of this, Binder looked back on the sale of the *Scheersberg* with vague uneasiness: It really had taken place a little too quickly, and Bolten still knew nothing whatsoever about Yarisal or his background. It crossed his mind that the ship was perhaps going to be used for gun-running, 'or some other shady business.'

Chapter 6

It's Okay By Euratom

Sailors say that a new ship, like a new woman, takes a bit of getting used to, and it is customary when a ship changes hands for the fresh crew to learn about her quirks and foibles from the men they are replacing. For this reason Captain Ludwig Hansen, master of the *Scheersberg*, stayed with the ship in Rotterdam after her drydock to await the arrival of the new owner's captain. So, too, did the *Scheersberg*'s chief engineer, Rudolf Kreuzfeldt, who had considerable experience of the ship's engine to pass on. They need not have bothered. On 28 September the day after Yarisal bought the *Scheersberg*, Hansen received orders to abandon ship. He paid off Kreuzfeldt, along with the cook and steward who had also remained on board, and returned to his home in Lubeck, northeast of Hamburg. Hansen felt uneasy at leaving the ship unattended and more than a little puzzled by what had happened.

Hansen would not be the only one who failed to understand what Yarisal was up to in the last months of 1968, for the *Scheersberg* was to be involved in a number of odd events as Operation Plumbat neared its climax. But the explanation for what happened to Hansen, at least, is probably quite simple: his replacement as captain was not yet ready to come aboard. Yarisal had been preoccupied with other changes. One appears academic; the other, less so.

First, Yarisal gave the ship a new name, in the most economic fashion, by the addition of a single letter: the *Scheersberg* became the *Scheersberg A*. (Yarisal was fond of this device. His ship *Flower Bay* was previously named *Flower Boy*.) Second, he applied to remove the ship from the West German register and place her under Liberia's more convenient flag. This switch in registration brought Yarisal immediate benefits. Unlike West Germany, Liberia did not impose regulations about the make-up and number of ship's crew. Under her new flag the *Scheersberg A* could sail with as few hands as Yarisal cared to employ.

On 29 September, the day after Captain Hansen went home, a scratch crew of Spaniards, Portuguese and Moroccans was recruited in Rotterdam. They were joined soon afterwards by two officers who had previously served on Yarisal's ships the *Linora* and the *Almaflora.* On 2 October, still without benefit of either a captain or a chief engineer, the *Scheersberg A* left Rotterdam for a one-day voyage along the north Dutch coast to the West German port of Emden.

It was only then that Yarisal set about making up the ship's complement. One man Yarisal wanted aboard was Peter Koerner, a German ship's engineer who had sailed for Yarisal on both the *Linora* and the *Almaflora.* Koerner had met Yarisal a couple of times, and once even dined with him. He shared the judgment of those who held Yarisal to be a gentleman. He admired his courtesy and his restrained good manners, and rather hoped that Yarisal thought the same of him.

Koerner was at home in Hamburg when Yarisal telephoned him. Yarisal told Koerner he had acquired a new ship, and joked about the 'lucky coincidence' that, like the previous vessles Koerner had served on, the name of this ship also ended in the letter A. Yarisal asked Koerner to become her chief engineer. Koerner accepted and agreed to take the train to Emden as soon as possible.

Koerner arrived at the port on 7 October. The *Scheersberg A,* he found, was not the most elegant of ships. She was one of the sea's workhorses, built strictly to purpose. Her twin holds occupied almost three-quarters of her length, with derricks on the bow and amidships and a gantry crane in front of the ship's rather cramped bridge. The crew's quarters, he discovered, were not exactly spacious, either.

Koerner's opinion of the *Scheersberg A* fell a little further when he made a closer inspection of the ship. Her paint work had been neglected and both her decks and hull were scarred with rust. Of particular concern to Koerner, however, was what he found when he clambered down the narrow stairway to the engine room. The engine appeared to have been deprived of regular maintenance.

Koerner would have liked to strip the engine immediately, but there was no time. The *Scheersberg A* had a bulk cargo to carry from Emden to Naples in Southern Italy and she was due to sail just as soon as her new captain turned up.

The captain arrived shortly after Koerner. He said that he was a Welshman, from Cardiff. He was young for a skipper, perhaps thirty-five, but he was very assured, very professional. He had a predilection for dark blue shirts and dark blue trousers, like a navy man. He said his name was Peter Barrow.

The *Scheersberg A* began her passage south to Naples on 9 October. The crew spent the voyage, and the rest of October, cleaning her up, while Koerner overhauled the engine as best he could. Barrow stalked the ship relentlessly, learning everything about her there was to know.

The crew did not know it, of course, but the long trek south from northern Europe, down the Iberian peninsula, through the Strait of Gibraltar and into the Mediterranean was a rehearsal for a voyage to come. And while Barrow was familiarizing himself with the *Scheersberg A*, other players in the drama were securing her future cargo.

Back in the Brussels office of SGM, Denis Dewez fulfilled the promise he had made to Herbert Schulzen in Wiesbaden. He drew up a formal contract for the uranium deal. It stipulated that SGM would supply Asmara Chemie with 200 tons of uranium oxide for a price of 190 million Belgian francs – about £1.5 million. Herbert Schulzen signed the contract on behalf of Asmara on 10 October. Dewez then set about the business of winning approval for the deal from Euratom, the Common Market's nuclear watchdog. It was not a difficult task.

Euratom was divided into two halves. There was a supply division, based in Brussels, which had the job of acquiring uranium supplies for countries which belonged to the EEC, and there was a safeguards division, then in the process of moving from Belgium to Luxembourg, which was supposed to monitor movements of uranium and spot any diversions. Together they were charged with controlling the use of nuclear materials in Europe, but they laboured under appalling limitations.

Euratom was formed at the same time as the EEC in 1957 in the hope that it would contribute to the political unification of Europe. That hope was soon dispelled. Euratom could only work if the member countries surrendered some of their sovereignty for the greater good of the whole. They were not prepared to do so.

Squabbles continually erupted among the six members about what role Euratom ought to play. It was above all France, the EEC's leading nuclear nation, who did her best to wreck Euratom, treating it with ill-concealed contempt. In 1962 France managed to install Pierre Chatenet as Euratom's president. Chatenet announced that Euratom should limit itself to collecting and centralizing information. In 1963 France submitted a memorandum complaining that Euratom was spendthrift and overambitious, and voted against the next Euratom budget. Any controls Euratom tried to enforce unfailingly violated somebody's national rights and were accordingly ignored. In December 1967 Euratom, torn by constant dissent, almost ceased to exist. The six members failed to agree on its next five-year plan and only after vigorous efforts by West Germany did they vote to continue Euratom's budget, and thus the agency itself, on a year-to-year basis. By 1968 Euratom was a cripple.

One curious result of the dissension inside Euratom was that the safeguards people had no say over deals such as the one that SGM and Asmara had in mind. What responsibility there was for vetting such transactions lay with the supply division, and in particular with Felix Oboussier, a personable German who was not by training a nuclear expert, but a lawyer.

Having secured Schulzen's signature on the uranium contract, Dewez paid Oboussier an informal visit. Even for a company the size of SGM, a £1.5 million sale was significant, and Dewez took to Oboussier's office some support in the person of Jean Mikolajcak, titular head of the uranium division. Mikolajcak carried with him the prestige of being the son of SGM's founder.

Dewez and Mikolajcak gave Oboussier a rough outline of what Asmara had said was going to happen to the 200 tons of uranium: it would be shipped from the Belgian port of Antwerp to Genoa, Italy, and then be taken to SAICA's plant near Milan; after treatment it would be shipped back to northern Europe to the Dutch port of Rotterdam, and then transported by road or rail to Asmara in Wiesbaden; there it would be used for what is known in the jargon as non-nuclear purposes – namely, as a chemical catalyst. Oboussier could see nothing objectionable in the scheme even though the uranium would technically leave Euratom's jurisdiction for the length of the voyages to and from Italy. To be on the safe side, however, he said that he would like to have a letter from Asmara, confirming the details.

Dewez and Mikolajcak agreed to pass on the message to Asmara. Just before they left, they mentioned to Oboussier that the Asmara deal was of course extremely confidential. Oboussier was used to that. Everybody who came to see him about uranium deals wanted the details kept secret; it was that kind of business.

A few days later, on 17 October, Schulzen wrote to Euratom on Asmara's behalf confirming the account Dewez had given. On reflection, one part of the tale did trouble Oboussier a little, and that was Asmara's claim that it was going to use the uranium as a catalyst. That was a practice Oboussier had never heard of. Prudently he decided to check its feasibility with some of the more technically minded members of Euratom's staff. The reply was reassuring. As it happened, a month earlier the Dutch government had bought some uranium for the very same purpose. That lucky coincidence dispelled any remaining doubts. Nobody in Euratom saw any need to make inquiries into Asmara's background, or, indeed, into any other aspect of the deal.

On 22 October SGM sent a hand-delivered letter formally applying for Euratom's sanction. Under its own regulations Euratom had eight days in which to raise objections. It did not. By Euratom's silence, on 30 October the sale of the uranium and its shipment by sea from Antwerp received automatic official approval.

On that date, the *Scheersberg A* was heading north from Casablanca towards the small port of La Pallice on the west coast of France. On 6 November she left La Pallice and, her pilot run to Naples and back nearly completed, headed north once again, bound for Rotterdam and Antwerp.

Mossad's luck had held so far. Israel all but had her uranium. What remained was to get it home.

Israel's uranium was waiting to be collected from the storage silo at the giant metal refinery close to the village of Olen. It was in the form of uranium oxide, known as yellowcake, which has the colour of an egg yolk and the texture of coarse sand. Moving 200 tons of the stuff the 4,000 miles from Belgium to Israel obviously presented considerable logistical problems. Nowhere were there more complications than over the first leg of that journey, the twenty-five miles between Olen and the port of Antwerp. That part of the trip would be overland, and transporting the uranium would need special care, given its terrifying propensities.

Uranium oxide was discovered in 1789 by the German physicist Martin Klaproth who named it in honour of the planet Uranus, discovered just eight years earlier. For a long time uranium was studied as a mere scientific curiosity since no practical value for it could be found. It was through his experiments on uranium ore that the French physicist Henri Becquerel discovered radioactivity in 1896. Not long afterwards the French husband and wife team Pierre and Marie Curie discovered that the most radioactive element in uranium ore was radium. They did not realize just how deadly it was, but were saved from serious ill effect at the time by the winds blowing through the drafty attic their poverty forced them to use as a laboratory. They later died of radiation sickness. So, too, did the watch girls of New Jersey, employed in the 1920's to paint radium on watch faces to make them luminous. They licked their paint brushes and absorbed radium into their bodies. Their doctors, who did not know of this new occupational disease, said they must have died from angina or syphilis.

Uranium was soon to become the most controversial of all the elements. In 1938 two German physicists discovered that when bombarded with neutron particles the uranium atom splits. They realized that if a succession of atoms could be split in a chain reaction, staggering amounts of energy would be produced. One pound of uranium, it was later shown, could yield as much energy as three million pounds of coal. The first self-sustaining nuclear chain reaction was conducted by Enrico Fermi at the University of Chicago on 2 December 1942. The success of that experiment led inevitably to the atomic bomb. The first was detonated on 16 July 1945, at a test site at Alamagordo Air Base in the New Mexico desert. Three weeks later the United States Air Force dropped atomic bombs on the Japanese cities of Hiroshima and Nagasaki. One hundred and forty thousand people died.

Yellowcake in its raw form, however, is relatively harmless. But it is still radioactive, and dangerous to swallow or inhale. As one British atomic energy official said: 'You wouldn't want to put your bed on it.' Not surprisingly then, governments and local authorities have imposed considerable restrictions on the manner of its transportation. The journey from Olen to Antwerp was not going to be simple. Fortunately for Mossad some very accomplished help was available, and it was now unwittingly recruited.

As Belgium's biggest port, Antwerp was equipped with several major freight companies. Any one of them could cope with the technical difficulties of shifting 200 tons of radioactive material the short distance to Antwerp. But it would not be easy to hire one of these companies. Because of the occupational hazard of becoming involved in hauling illicit cargo, they were very cautious about who they did business with. Since they were so fussy it was necessary to approach one of them by a circuitous route.

Towards the end of October Asmara hired a Frankfurt broker named Klaus Rehwagen who had considerable contacts in the freight business. Rehwagen in turn approached the Belgian transport firm of Ziegler & Company, with whom he had done business in the past. Ziegler happily accepted the assignment because it knew and trusted Rehwagen. Ziegler made only one check on Asmara itself, into its credit rating. The First National City Bank in Amsterdam reported it was satisfactory.

Once hired, Ziegler provided a very comprehensive service. The job was assigned to Ziegler's Antwerp manager, Marcel Heynen, and he enthusiastically found out about all the formalities involved in moving the uranium. He obtained the necessary permissions from the Belgian Ministry of Transport and the Belgian police. He hired a train to carry the goods from Olen to Antwerp. He hired a berth in Antwerp dock where the ship could be loaded. He hired stevedores and cranes to do the work. And he hired river pilots and dock pilots to get the boat in and out of the harbour. Asmara, and the men behind it, did not have to expose themselves to any further official scrutiny.

During this feverish period Heynen frequently telephoned Schulzen to let him know how things were going. At the end of the first week of November Heynen called him again to say that Ziegler was ready to move. He asked Schulzen for the name of the ship that would carry the uranium to Genoa, and for final confirmation of the date when she would be in Antwerp. The date was 16 November. Schulzen said that the name of the ship was the *Scheersberg A.*

In view of all the hard work they had put in to cleaning her up, it was with some dismay that the crew of the *Scheersberg A* learned early in November that they were to be discharged. Captain Barrow broke the news as the little ship headed north back to Rotterdam, barely a month since she had left Emden for Naples. He said that Yarisal had sold the ship and that the new owners would be supplying their own crew.

The *Scheersberg A* arrived in Rotterdam on 11 November. Peter Koerner, like his predecessor, expected that he would pass on his knowledge of the engine to the replacement chief engineer. Like his predecessor he was disillusioned. Barrow said his orders were that the entire crew, himself included, should be paid off immediately. They gathered their belongings and departed the same day, leaving the ship unattended.

Four days later, on 15 November, the *Scheersberg A* left Rotterdam, bound for Antwerp. Her new crew was unremarkable except it was half the normal strength and the seven deck hands were all Caucasian. The captain was an assured, professional young skipper with a predilection for dark blue shirts and trousers, just like a navy man. He said his name was Peter Barrow.

Chapter 7
Operation Plumbat

November 15, 1968. At Olen 200 tons of uranium oxide was removed from the silo. It was packed in 200-litre oil drums which were closed with lids held in place by substantial rubber seals. The 560 drums carried nothing to show what they now contained, except for the word PLUMBAT stenciled on each lid at Asmara's request.

The railway tracks at Olen run beyond the village station and directly into the plant. Eleven covered wagons that had been hired by Ziegler, together with an engine to haul them, were shunted alongside the silo. The drums of PLUMBAT were loaded on to the wagons, which were then locked. Two guards, one provided by Ziegler, the other by the Belgian railroad company which owned the wagons, climbed on board. Under their watchful eyes the train set out on its short journey to Antwerp.

While it was on its way Herbert Schulzen turned up unannounced at Ziegler's office in the heart of the Antwerp docks. He was accompanied by a short fat man wearing spectacles and the two of them fussed about last minute details. Ziegler's manager Marcel Heynen assured them that the arrangements were running like clockwork.

That happy state of affairs continued. Just after four o'clock the next morning the *Scheersberg A* arrived at Flushing at the mouth of the Schelde, the river which runs from Antwerp to the North Sea. A pilot was waiting to guide the ship the last forty-five miles to the docks. He went on board. In Antwerp, Ziegler's water clerk Josef Verhulst was woken at home by a telephone call from Flushing, alerting him that the *Scheersberg A* was on her way. Verhulst calculated that loading the uranium could begin early that afternoon. He dressed and went to his office to make the final preparations.

At 8 am the *Scheersberg A* arrived at the entrance to Antwerp docks. She still had some way to go to her berth and the river pilot who had guided her up the Schelde was joined by a dock pilot. His task was to negotiate a safe passage through the system of locks that led to the berth. He did not actually steer the ship; that was traditionally the job of the first officer.

The first officer of the *Scheersberg A* was, according to his seaman's papers, an Englishman named Tilney. He had been given precious little time to get the feel of the ship but he took her helm nonetheless. The dock pilot stood at his shoulder giving guidance while Captain Barrow watched in silence. There was no time for small talk. Strong gusts of wind were pushing at the *Scheersberg A* and even with the help of an attendant tug boat, Tilney had difficulty in holding her on course.

Leaving the main harbour channel the *Scheersberg A* turned left into Royers Lock. She passed safely through both sets of lock gates and then turned sharply to starboard to approach the entrance to a little canal named Suez. At the entrance to the canal was a lifting bridge with the even more unlikely name of Siberia. Tilney had to line up the *Scheersberg A*'s bows to pass under Siberia bridge but as he edged her forward, nature played a malevolent trick.

A maverick gust of wind caught the ship's bow and swung it towards the bridge's bluestone support. It seemed that the *Scheersberg A* must ram the support, wreck the bridge, and block the canal. But at the last moment the gust died and the *Scheersberg A* swung away. Even so her starboard hull grazed the bridge. On examination, the damage proved superficial and the port authority accepted Ziegler's promise that the bill, for a few thousand Belgian francs, would be paid.

The emergency over, the *Scheersberg A* passed through Suez into the Kattendijk Dock and tied up at berth 42. It was now just after noon. The railway wagons bearing the valuable cargo were already waiting on tracks that ran right to the water's edge. Also waiting for the *Scheersberg A* were Ziegler's water clerk Josef Verhulst, and Herbert Schulzen's companion, the short fat man wearing spectacles. Both went on board.

It was Verhulst's job to complete all the formalities that attend the arrival in Antwerp of any ship. He first paid the two pilots their customary gratuity. Then he asked Captain Barrow for a list of the ship's crew. He passed it on immediately to the harbour police but none of the names on it excited any curiosity.

A 2 pm Ziegler's stevedores began loading the cargo. The drums of PLUMBAT were hoisted by cranes and lowered into the *Scheersberg A*'s two holds. There they were carefully secured with wedges of old timber.

In the early evening, with the job nearly done, Ziegler's manager Marcel Heynen went on board to make sure that everything had gone smoothly. He noticed the short fat man with spectacles counting the drums as they were loaded. He found Captain Barrow in his cabin: Barrow poured him a glass of Scotch whisky and the two men talked in English about the weather. It looked as though it was going to be fair. Heynen wished Barrow bon voyage and went ashore.

By nightfall loading was complete. Water clerk Verhulst went on to the *Scheersberg A* to obtain a signature for his records. He needed one of the officers to sign a receipt stating that 208,998.55 kilos of Plumbat (the weight included the metal drums) had been 'safely received'. First Officer Tilney was willing to sign but only after he had added in ballpoint pen the caveat that the cargo consisted of 'drums said to contain chemicals.'

At 10 pm the *Scheersberg A* was ready to sail. A new pilot, Pierre Slaghmeulen, was on board to guide her out of the docks. Verhulst telephoned customs and the police to obtain their clearance. Customs and police officers went to Berth 42 and boarded the ship, but their only interest was in contraband and stowaways. Having found neither, and having received their customary gratuities, they left.

At four minutes past midnight, on Sunday 17 November, the *Scheersberg A* sailed. Eight hours later she reached Flushing at the mouth of the river, discharged her pilot and set out across a cold forbidding sea.

* * *

The uranium had only been allowed to leave Belgium and the boundaries of the European community on the strength of Asmara's promise that it was going to Italy. For the first week the course of the *Scheersberg A* did nothing to damn that lie. She sailed southwest down the coast of Belgium, through the Strait of Dover and into the English Channel, the most congested stretch of ocean in the world. For the first two days the little ship battled through rough seas and strong winds that blew from the southeast, sometimes gusting to gale force. But in the early evening of 19 November, as she approached the westernmost tip of France and the Atlantic ocean, her luck changed. The wind moderated then died. For the remaining three days that it took the *Scheersberg A* to reach the Strait of Gibraltar and the Mediterranean, the seas she sailed were a flat brooding calm.

At around midday on 24 November she reached the point of no return. Having passed the Spanish Balearic islands she should have turned north for Genoa. Instead, she kept going east. On 27 November she passed between Tunisia and Sicily and into the eastern Mediterranean.

The *Scheersberg A* could not take the uranium directly to Israel without giving the game away. Her arrival would have been routinely reported to Lloyd's of London which maintains agents at every port and keeps a record of shipping movements. So, in halcyon weather, she headed for the waters between Cyprus and the coast of Turkey – and for a silent rendezvous with an Israeli freighter.

According to an anonymous seaman* who claims to have been on board the *Scheersberg A*, the rendezvous was set for the early hours of a December morning. An Israeli freighter, escorted by two gunboats, fiercely armed, approached from the south. The freighter manoeuvred alongside the *Scheersberg A* until lines thrown between the ships could hold the two together. In the holds of the *Scheersberg A* – one amidships, one forward – the ship's seven deckhands removed the timber wedges that secured the precious cargo. The 560

***Time* magazine says that one of its sources spoke to the seaman in the Ivory Coast in 1973. (*Time*, 30 May 1977).

metal drums were winched one by one through the ship's hatches, lifted high above her deck, and swung carefully onto the freighter. On board the *Scheersberg A* no one spoke; Captain Barrow had ordered complete silence. But once or twice the crew heard a sharp command in Hebrew, coming from the decks of the freighter.

The transfer must have taken about four hours. When it was over the lines were freed and the freighter, now 200 tons heavier, slipped away on a course taking her almost due south, back the way she had come. Her escorts followed. When they had gone, Barrow turned the *Scheersberg A* northeast, towards the Turkish coast. Sailing at twelve knots, she would have reached the little port of Iskenderum in less than five hours. She arrived there early on 2 December.

The Israeli freighter had a longer voyage. It would have taken her perhaps sixteen hours to reach the Israeli port of Haifa. From then on there was little need for secrecy. The metal drums, labelled innocently PLUMBAT, could have been unloaded quite openly and carried south on the main Tel Aviv road by a small convoy of lorries. By nightfall on 3 December the uranium would have reached its final destination. Dimona.

A few days later, Peter Koerner, at home in Hamburg, got yet another telephone call from Burham Yarisal. He said he wanted Koerner to meet the *Scheersberg A* in Palermo, Sicily, where she was lying unattended. Koerner was surprised. It was less than a month since he had been discharged on the grounds that Yarisal had sold the ship. 'Ah well,' said Yarisal, 'I've bought her back.'

On Yarisal's instructions, Koerner met up with a colleague. The two men collected rail tickets that were waiting for them at the local office of the Hapag-Lloyd travel agency and took the long train ride south to the toe of Italy where they crossed by ferry to Sicily. They arrived in Palermo on 11 December to find they were not the only ones who had been summoned by Yarisal. All the old crew of the *Scheersberg A* were there – all that is except for Captain Barrow.

There was a new skipper, a Spaniard named Francisco Cousillas. He shared the crew's curiosity about what had happened to the *Scheersberg A* during the past month. They turned to the ship's log to find out where she had been since Rotterdam.

The last two pages had been torn out.

Operation Plumbat was so beautifully conceived and executed that it was months before the European authorities even realized that the uranium had gone astray. And, as we shall see, the investigations that were launched did not uncover a sliver of evidence as to where the 200 tons of yellowcake had gone. It was not until almost five years later than Dan Aerbel, in a prison cell in Norway, provided the evidence by blurting out the name of the *Scheersberg A*.

Aerbel's role in Plumbat had been to dispose of the evidence – to get rid of the *Scheersberg A* after her voyage, without leaving any clues that could link the ship to Israel. It was not in theory a difficult task. But, perhaps seduced by the brilliance of its success, Mossad decided to use the *Scheersberg A* for one further audacious operation before letting her go. This escapade came close to disaster and, as a result, when Aerbel was finally instructed to sell the ship he did so in a panic, scattering documentary evidence.

As with Plumbat, it was Inspector Steinar Ravlo of Norway's E-Gruppa who got a confession from Aerbel about the second mission for which Mossad used the *Scheersberg A*. This time, however, it was an enterprising Danish journalist named Ole Schierbeck who really deserved the credit.

Schierbeck had taken a keen interest in the Lillehammer murder ever since learning that one of those arrested, Aerbel, came originally from Denmark. He spent several weeks researching Aerbel's background, talking to more than one hundred people; and he discovered, among other things, that Aerbel had once owned the *Scheersberg A*. Schierbeck did not learn about Operation Plumbat because the European authorities, deeply embarrassed by the affair, had deliberately covered up the loss of the uranium. But he did appreciate that there must have been a good reason for a Mossad agent to own a cargo ship. From Lloyd's Shipping Register in London Schierbeck obtained a list of the *Scheersberg A*'s movements for 1968 and 1969. And as he studied the dates and places it contained a sudden chord was struck in the recesses of his memory. He recalled the strange story of Knud Lindholm Pedersen.

In October 1970 Pedersen, a veteran Danish sea captain aged fifty-two, was arrested in England. He had been attempting to land a group of illegal Asian immigrants when his ship, a former German torpedo boat said to have once been used for a pleasure cruise by Hermann Goering, broke down. Pedersen's boat was towed into Harwich harbour and he was held by police. He appeared in court, was bailed, and promptly fled to Holland. The British police followed him but were powerless to act as Pedersen had not been charged with an extraditable offence. A full account of these events appeared in Schierbeck's newspaper *Politiken.*

In March 1971, Schierbeck received an anonymous letter written on stationery headed 'Hotel International, Hamburg.' The letter complained that the stories *Politiken* had published about Pedersen were full of errors. Schierbeck had not written the stories, but he called the Hotel International in Hamburg. The phone was answered by Pedersen. He suggested that Schierbeck come to Hamburg to meet him. Schierbeck caught a plane from Copenhagen that evening.

Schierbeck found that the Hotel International was a less than glamorous sailors' hotel on the Hamburg waterfront. Pedersen was waiting for him and they went up to a dormitory and opened a bottle of whisky. Pedersen proceeded to tell Schierbeck the sordid facts involved in smuggling human beings.

It was already after midnight when Pedersen finished his account. Schierbeck casually asked him what else he had been up to.

'Do you remember the Cherbourg affair?' Pedersen asked him. Schierbeck said that he did.

'That was me.'

Pedersen then told an astonished Schierbeck that he had helped the 'Israeli secret service' to plan one of its most extraordinary operations – the 'liberation', in December 1969, of five gunboats from the French port of Cherbourg. Pedersen claimed that he had navigated one of the escaping gunboats on its hazardous route out of Cherbourg harbour. He also described the refuelling of the gunboats by a cargo ship, '300 sea miles southwest of Lorient' in the Bay of Biscay. Pedersen coped ably with every question Schierbeck fired at him. He did not ask Schierbeck for money for the story, and by the time he took his leave Schierbeck was convinced that Pedersen had been involved in the operation in some way. Back in Copenhagen, Schierbeck wrote the story as Pedersen told it, and *Politiken* printed it. Schierbeck was reluctant to believe that in its length and richness of detail Pedersen's account could be total fabrication, but where truth ended and fantasy took over, he could not decide.

Now, in August 1973, the whole episode came vividly to Schierbeck's mind. And when he studied the movements of the *Scheersberg A*, as logged by Lloyd's, he saw how her course precisely intersected with the route the gunboats must have followed from France to Israel. Schierbeck was certain he had discovered the reason why Mossad's Dan Aerbel had once owned a ship: the *Scheersberg A* was surely the refuelling vessel Pedersen had described.

Schierbeck knew it was impossible to check with Pedersen. Nine months after their meeting Pedersen had died in a mysterious accident on board a ship in Copenhagen harbour. Nonetheless, Schierbeck wrote a story for *Politiken* speculating that the *Scheersberg A* – and therefore Aerbel – had played a vital role in the 'liberation' of the gunboats.

In Norway Inspector Ravlo learned of Schierbeck's story. He asked Aerbel if it was true. Aerbel said it was, and provided what details he could.

Once again he supplied the final piece of the jigsaw. Looking at the complete picture it is striking how much of a model Operation Plumbat was for the notorious gunboats affair.

Chapter 8
Gunboat Diplomacy

The gunboats affair: the phrase itself has an intimately French ring. It was one of those dramas in which France seems to specialize. There were the thunderous pronouncements and accusations of betrayal made by politicians; the contradictions and ambivalences of state policy laid bare; and the obvious enjoyment of the people of Cherbourg, in their role as chorus, at their leaders' discomfort. For Israel the episode was manna: one more audacious chapter to be written in her breathless and valiant history. Israel naturally played up the elements of personal courage. But what had preceded the events of Christmas Eve 1969 were many months of detailed planning and meticulous preparation. It was almost as if Mossad, after Plumbat, had agreed to an encore.

The Cherbourg affair had its origins in the last days of 1968, barely a week or so after Israel had successfully brought her uranium home. On 26 December an El Al Boeing 707 crammed with passengers on the long journey from Tel Aviv to New York sat on the tarmac at Athens airport, ready to resume flight. Suddenly its fuselage was raked by gunfire, hand grenades were hurled at its engines and fuel tanks. The airliner had been chosen as a target by two young Palestinian terrorists. One passenger died and the others were gripped by fear but miraculously the holocaust the terrorists had hoped for did not ignite. The two men fled to the airport terminal where they were seized by Greek security police. They said they belonged to the Popular Front for the Liberation of Palestine, and they had flown to Athens from Beirut.

Two days later Israel took her revenge. If Lebanon harboured terrorists, Defence Minister Moshe Dayan decided, then Lebanon must pay. On 28 December Israeli commandos landed in helicopters at Beirut's international airport. They draped explosives over thirteen empty passenger planes belonging to Middle East Airlines and detonated them. As the helicopters took off again for Israel, the blazing planes lit the sky. The destruction totalled £20 million.

General Charles de Gaulle, President of France, learned of the raid as he watched the television news at his home in the village of Colombey-les-deux-Eglises, where he was spending the Christmas holidays. Israel's attack offended him deeply. He had fond memories of his service as a young army officer in Lebanon in the days of French rule, and he felt that his personal intervention had managed to save Lebanon from being caught up in the Six Day War. Now the Israelis had landed in French-made helicopters to destroy French-made planes belonging to an airline in which Air France was a major shareholder. On New Year's Day, de Gaulle made a speech at a formal diplomatic reception at which the Israeli ambassador was present. He attacked Israel's aggression towards 'a peaceful country and a traditional friend of France.' And he decreed that from now on France would supply no more arms to Israel.

It was not the first time de Gaulle had issued such orders. He had made the same threat at the time of the Six Day War eighteen months earlier. His main target then had been the fifty Mirage jet fighters Israel had ordered from the French Dassault company. Israel was prevented from taking delivery of the Mirages but otherwise the embargo had been an ambivalent and hypocritical affair, reflecting as much as anything the ambiguities of French policy-making under de Gaulle. But this time the General made it clear that he meant what he said. The most serious casualty of his new, tough attitude were the missile-launching gunboats being built for Israel in the dockyards at the French port of Cherbourg.

Any navy would be proud to own the gunboats. Just under 150 feet long, they were designed to reach a top speed of more than forty knots, yet, under Israel's specification, pack the punch of a destroyer. They were to be armed with torpedos and rapid-fire guns, and most formidable of all

Gabriel missiles, designed and built by Israel herself. The missiles could deliver a 330 pound warhead with fearsome accuracy to an enemy ship twelve miles away, thanks to a highly sophisticated guidance system which would take it skimming over the waves. The Israeli Navy had first asked a German shipyard to build the gunboats, but when Germany stopped giving Israel arms the order was switched to the Cherbourg company Constructions Mecaniques de Normandie, known as CMN. Israel ordered six gunboats on 26 July 1965, and six more on 14 March 1966. The price was £700,000 each.

The first boat was launched on 11 April 1967. Six months later Israel's urgent need for the boats was confirmed when her destroyer the *Eilat* was sunk off Port Said by a Styx missile launched from a Soviet-built Komar missile boat of the Egyptian Navy. That disaster left the Israeli Navy with one old destroyer, six submarines, a few patrol boats and landing craft. Egypt would rule the waves until the French gunboats were delivered. De Gaulle's first embargo, imposed after the Six Day War, did not affect the gunboats and the first five reached Israel safely. Then came Beirut, and the General's rage. He did not even pay Israel the courtesy of notifying her formally of the new policy, merely instructing the French customs to prevent any more gunboats from leaving Cherbourg.

Israel did, however, receive advance warning of the ban from a sympathizer in the French Ministry of Defence. As a result she was able to get boats six and seven out of Cherbourg just in time. The sixth boat had conveniently left Cherbourg for sea trials on 31 December, and the following evening her commander radioed that everything was in order and he would sail directly to Haifa. He went only as far as Gibraltar. The seventh boat left Cherbourg harbour on 4 January, just as the new embargo instructions were clattering in on the telex machine in the local customs post. Boat number seven met boat number six at Gibraltar and the two sailed home together.

They were the last to escape. When gunboats eight and nine were launched early in 1969 they remained securely in Cherbourg. Previous boats had been guarded in the naval section of the harbour but after the new embargo was imposed Cherbourg's *prefet maritime* refused to take delivery of any more. The two new ones suffered the indignity of being moored to the pier used by the ferries bringing tourists and their cars from England.

In April 1969 General de Gaulle was approaching the end of his eleventh year as President of France. He had consolidated his power through referendums, and having survived the dramas of May 1968, when students fought riot police nightly in the streets of Paris and France was swept by strikes, he now decided to use a referendum for the fifth time. The question he posed was somewhat abstruse, concerning a reform of the constitution in order to alter the system of regional government. It is said that governments offer referendums only when they are certain what the answer will be. If that is so, de Gaulle miscalculated. He lost. He resigned immediately and returned in pique to Colombey-les-deux-Eglises. On 15 June France elected Georges Pompidou her new President.

The question of the Israeli arms embargo arose at President Pompidou's first press conference. He was cautious. He spoke of the embargo possibly becoming 'more selective' at some future date, but for the moment it would remain in force. In private Israeli diplomats pleaded with their counterparts at the French Foreign Ministry to relax the ban and release the gunboats, but they received only the same irritating half promises that the situation might perhaps be reviewed one day – but not yet. While the manoeuvring continued, gunboat number ten was launched on June 30 to join her two sisters in idleness at the car-ferry terminal.

Admiral Mordecai Limon had no time for the nuances and posturing of diplomatic life. Once a Navy man and now Israel's chief arms buyer in France, he was in no doubt what should be done: Israel should seize the gunboats. It was not surprising that he should hold so forthright a view, given his career as an Israeli patriot and man of action. Born in Poland in 1921, he had been taken to Palestine as a young child in the second great wave of Jewish immigration. In 1940 Limon had captained small boats that, with Mossad's help, landed illegal Jewish immigrants on the beaches of Palestine. In 1948 he was made an admiral in the new Israeli Navy and in 1950, at the age of thirty-one, he became its Commander-in-Chief.

In 1954 Limon was chosen for a less public role. He was sent to New York, then brought back to Tel Aviv where he took up a series of posts in Israel's Ministry of Defence. Even their formal titles convey their sensitivity: deputy director-general in the department of 'emergency economic planning'; assistant director-general in charge of 'special duties'. In 1962 he was transferred to Paris, which Mossad had chosen as the European headquarters for its intelligence gathering and undercover operations. Limon was made a 'special envoy' at the Israeli Embassy, where his assignment included supervising Israel's military purchases. He had been engaged on the Navy's gunboat order since 1965 and had become increasingly frustrated by France's prevarication. Finally Limon flew to Tel Aviv. He met Defence Minister Moshe Dayan and argued that Israel must take the boats. Dayan agreed.

The decision had been quickly made; putting it into effect was quite another matter. As it pondered the best way of taking the gunboats, the Israeli Cabinet called in Mossad.

The head of Mossad, General Zwi Zamir, had two broad possibilities before him. The gunboats could be hijacked through surprise and the force of arms. Or they could be spirited away in a clandestine operation. The first rather dramatic approach accorded more with the traditions of the Israeli Navy and the preferences of the impatient Admiral Limon, and Zamir agreed to evaluate its chances of success.

A senior Navy man, Rear Admiral Benny Telem, had already been to France to look at the potential scene of action. He went ostensibly to take a member of his family for an eye operation. While there he travelled to Cherbourg and examined the harbour. It was certainly tempting, Telem reported. It was a great advantage that the gunboats were moored in the civil harbour, for it was lightly guarded, access to the quays was straightforward, and the open sea beckoned invitingly beyond two gaps in Cherbourg's long outer sea wall.

Now an occasional agent for Mossad made other soundings. In the German port of Cuxhaven he contacted the Danish sea captain Knud Pedersen – who was later to tell his story to the Danish journalist Ole Schierbeck. As Pedersen's experience included service on patrol boats for the British Navy during World War Two, and later the rather less honourable trade of smuggling Asian immigrants into Britain in a former German torpedo boat, there was some truth in his claim to be skilled in piloting small vessels in difficult conditions. What was more he claimed to have made a special study of Cherbourg harbour. He too argued in favour of snatching the Israeli gunboats under cover of bad weather or darkness.

It was ironic, given Mossad's past, that in Tel Aviv Zamir found himself arguing against these views and in favour of a stealthy approach. It was true that snatching the boats fitted well with Israel's traditions of valour and daring. But there were clear risks involved in such an operation. Pompidou was no de Gaulle. But would even Pompidou allow so flagrant an insult to go unanswered? From Cherbourg to Haifa was more than 3,000 nautical miles. It would take the gunboats the best part of three days to reach the Mediterranean, and the French would have ample time to dispatch an intercepting force from her giant naval base at Toulon, near Marseille.

There was one striking case history to illustrate an argument for an operation based on cunning. That was Plumbat. Zamir had been in charge then too, and knew that Mossad could use 'friends' and front companies in conveniently far away places to get the boats, just as it had used 'friends' and front companies to get the uranium. And if that *modus operandi* were used for Cherbourg, there could be one exquisite bonus.

Mossad had just learned that France was secretly trying to sell Mirage jet fighters to Israel's enemy, Libya. They were the very same Mirage fighters that France had withheld from Israel under de Gaulle's first embargo. If Mossad could make it look as though France had been her accomplice in releasing the boats, then the negotiations with Libya could be badly damaged – perhaps wrecked. To Mossad, such piquancies have special appeal, and if this ploy came off it would more than compensate for the failure to make it appear that the PLUMBAT uranium had been bound for the Arab world when it disappeared. The Israeli cabinet duly gave permission for Mossad to lift the gunboats by stealth.

The Navy accepted the decision. But there was one practical problem that no amount of cunning could guarantee to solve. How would the gunboats obtain enough diesel fuel for the voyage from Cherbourg to Haifa? The precise distance was 3,090 nautical miles. The range of the gunboats depended on their speed. At fifteen knots they could travel 2,500 miles; at twenty knots, 1,600 miles; at thirty knots, only 1,000 miles. On the one hand there would be a need to save fuel; on the other, a desire to move fast enough to deter would-be pursuers. The gunboats' customary cruising speed of twenty-two or twenty-three knots seemed about right, which would give them a range of some 1,500 miles; although the possibility of bad weather also had to be weighed, as that would mean higher fuel consumption. All in all, it seemed clear that two refuelling stops would be necessary.

The earlier gunboats had taken on fuel in the ports of Gibraltar and Palermo, but it was unlikely that the escaping gunboats would be able to do the same. Refuelling at sea was the only alternative. Here the Navy was in difficulty. The Israeli fleet consisted of submarines, gunboats and landing craft – certainly nothing the Navy could send the length of the Mediterranean to meet the five boats on their way from Cherbourg.

Mossad had the answer. It told the Ministry of Defence it had one most suitable vessel available which had already proved satisfactory for the transfer of cargoes at sea.

Chapter 9
The Good Ship Sofa . . .

For most of 1969 the *Scheersberg A* had plied for hire like any normal cargo ship. Yarisal, having previously turned down the offer of the shipbrokers Bolten to find him cargoes, now asked the firm to do so after all. Bolten in turn overcame its earlier unease about Yarisal, and agreed. But the life of the *Scheersberg A* was not without complications.

On her very first voyage out of Palermo under Captain Cousillas, the *Scheersberg A* ran into a storm, the consequence of which did much to justify Peter Koerner's misgivings about the condition of the ship. In the violent movement, several of the bolts that held the engine's cylinder block sheered. Koerner had to carry out emergency repairs to hold the block until the ship could reach port and the bolts could be renewed.

There was a second emergency, in February, when she was on her way past Turkey – Yarisal's main base of operations – to the Roumanian port of Galatz on the Black Sea. In the Dardanelles, the narrow straits which divide the Mediterranean from the Black Sea – and also, it is said, Europe from Asia – the *Scheersberg A* ran aground. She strained several of her hull plates as a result and when she got to Rotterdam on 27 March she was put into dry dock for repairs.

The work was straightforward enough, but the incident did pose peculiar problems for the people behind the *Scheersberg A*. The work was to be carried out by a Rotterdam firm under the supervision of Germanischer Lloyd. As Biscayne Traders was a foreign company, Lloyd's needed the name of a local broker who would accept responsibility for the bill. The difficulty for Biscayne Traders was that it did not have a broker in Rotterdam. To secure the service of one would mean providing references, and that was not in Biscayne Traders' style.

The problem was neatly solved. Biscayne Traders managed to find someone in a broker's office who would forward the bill. Lloyd's were supplied with the name of the Rotterdam brokers Intermar – without that firm knowing its name was being used. The work was finished on 21 April and the bill was dispatched to the 'Captain and owners of the MV *Scheersberg A*, c/o Intermar NV, Willemskade 14, Rotterdam.' The amount owing, charged in Dutch guilders, was equivalent to £520, and it was duly paid. To this day Intermar insists that it has never heard of, let alone had any dealings with, either the Biscayne Traders Shipping Corporation or the man who was registered as its president, Burham Yarisal.

These incidents apart, the *Scheersberg A* sailed a conventional course. On leaving Rotterdam she embarked on her longest voyage of this period. She called first at the British North Sea port of Felixstowe to pick up a sailing yacht. With the yacht lashed to the top of her main deck she crossed the Atlantic to Boston, battling through several storms on the way. After unloading the yacht the *Scheersberg A* turned north, passing around Nova Scotia to reach the Canadian port of Newcastle in Miramichi Bay, New Brunswick. By the end of May she was back in Antwerp. She spent much of the summer in the Mediterranean, with the occasional diversion to Belgium and Holland, and one visit to a Polish port with the impossible name of Szczecin.

In early October the *Scheersberg A* was in the North Sea calling briefly at Antwerp again, and then putting in at Hamburg. There Captain Cousillas took on board a cargo that was bound for southern Spain and North Africa. The ship left Hamburg on 7 October.

It was now that Mossad decided the time had come to take closer control of the *Scheersberg A*. It is possible that Yarisal simply wanted to be rid of the ship; or perhaps the arrangement with him was proving rather expensive. Whatever the reason, Mossad could be forgiven if it was now feeling very confident. Almost a year had passed since the successful conclusion of Operation Plumbat. In all that time Euratom had come nowhere near tracing the uranium and 'the loss' had not even been made public. There was also the encouraging fact that the *Scheersberg A* had not been molested, though she had spent a good part of 1969 within easy reach of Europe's police forces.

In this favourable climate the first step was for one of Mossad's people to take over the affairs of Biscayne Traders Shipping Corporation. Dan Aerbel seemed the ideal candidate.

He had lived in Israel for eighteen months in the early 1960's but before and since then he had had no open links with the country. He had a Danish passport, and a range of names – they included Ert, Ertz and Erteschick – that he used in different circumstances. He had also been building his cover as an ambitious businessman who dealt increasingly in exports. For him now to purchase a ship was a perfectly logical step.

On the 'need to know' principle Aerbel was, at first, told nothing of the *Scheersberg A*'s role in Operation Plumbat. Indeed, at this juncture he was not told about the ship at all – only that he was to become the president of a Liberian shipping corporation. For Aerbel it was a painless process. On 16 October a nominee, who gave his name as Kurt Gellert, went before the notary public of Zurich to testify that Aerbel was 'entitled to sign on behalf of Biscayne Traders Shipping Corporation in all respects.'

The next day the *Scheersberg A* arrived in the North African port of Tunis. There Captain Cousillas received the startling news that once again the ship was to be sold. He was told little else. He gathered that the new owner was a Dane and he also understood that the ship was to be renamed. Even this, however, was a matter of some confusion because the radio message conveying the information had been broken during transmission. It should have read: 'So far as I know they don't have a name for it yet.' But it had been interrupted during the second word – 'So fa. . . .'

Captain Cousillas was told that the new owner wanted the *Scheersberg A* – or the *Sofa*, as the crew were now calling her – to go to the Swedish port of Lulea. It would mean a three week journey that would take the *Scheersberg A* to the very top of the Gulf of Bothnia between Sweden and Finland, within 100 miles of the Arctic Circle. Lulea is an important port for iron and timber but any business would have to be conducted briskly as the harbour usually freezes over at the end of November. At first light on 29 October the *Scheersberg A* left Tunis to start her long voyage north.

Chapter 10
A Little Help From Panama

With the *Scheersberg A* safely on course for Lulea, Mossad began its preparations for Cherbourg in earnest.

If France would not give Israel her gunboats Mossad would find someone France was willing to supply them to. Just as in Plumbat, Mossad now turned to its friends.

The first of these was the Israeli businessman Mila Brenner. He was president of Maritime Fruit Carriers, a large international corporation with offices in Tel Aviv and New York. He was well-connected, being married to a niece of Chaim Weizmann, Israel's first president; and he had a solid background of service to the nation, having fought as a ship's commander during the 1948 war and remained in the Israeli Navy until 1954. He had also been given the rank of 'reserve lieutenant general' in the Israeli army, a sure sign of official gratitude for services rendered. He had gone into business in 1957 and subsequently acquired other good reasons for feeling loyal to Israel. The prosperity of Maritime Fruit was based in no little part on its contract to ship Israel's renowned Jaffa oranges.

Brenner's job for Mossad was to construct the cover that would be needed for the fake switch in the gunboats' destination. For Plumbat, Mossad had taken advantage of the speed, convenience and informality Liberia offers the modern businessman. Brenner was more familiar with Panama, a country similar to Liberia in the services it can provide. For around £400, a company can be set up in Panama in a mere twelve hours. Brenner felt no inhibitions about his task. He went directly to Arias, Fabrega and Fabrega, one of the longest established and most respectable law firms in Panama.*

Brenner asked Arias, Fabrega and Fabrega to set up a Panamanian company, not an unusual request, and the formalities were soon under way. Just as in Liberia, instant companies are formed in Panama by proxy, and on 5 November 1969 the respected Senores Roy Carlos Durling, Fernando Cardoze, and Esteban Bernal became the founding officers of a new Panamanian company, incorporated under public deed number 2251 issued by notary public number five of the Circuit of Panama. The new company's name was Starboat.

As it goes about its work, Mossad likes to leave tiny clues to its participation; they are in the nature of private jokes. That sense of humour had led Mossad to mark the original destination of the Plumbat uranium as the Moroccan firm Chimagar, a joke that was only spoiled when Mossad discovered that the ensuing bureaucratic complications would be too great. In this case, the clue lay in the name Starboat.

The french word for a star (the show business kind) is 'vedette'. The French word for a gunboat (the Cherbourg kind) is also 'vedette'. The two meanings were elegantly combined to produce the word Starboat. The one nation that should have got the point perhaps was France. But then there was a great deal about the Cherbourg operation that the French did not appreciate until it was far too late.

Despite this tease, it would have been impossible for Mila Brenner to be named in the documents setting up the Starboat company. So Brenner in turn enlisted the help of one of his friends. Once again a lesson was drawn from Plumbat. Israel acquired her uranium by using Germans to buy it from Belgians and to ship it to an Italian in a boat owned by a Turk. Brenner now recruited the help of a Norwegian.

His name was Martin Siem. He and Brenner were old friends, and Brenner knew that Siem was privately sympathetic to Israel. More to the point, they had also done business together, and Brenner knew that Siem's cover would be immaculate. Siem was managing director of the giant Nor-

*Brenner would have been hard put to find any firm more respectable: its senior partner Hermodio Arias was Panama's president in 1940 and 1941; his brother Arnulfo was president from 1949 to 1951, and again briefly in 1968.

wegian Aker company, itself part of the Fred Olsen shipping and shipbuilding group, one of the cornerstones of the Norwegian economy. Aker was in the booming but fiercely competitive business of North Sea oil. Many European firms had found themselves shouldered out of the action by hard-bitten oilmen from Texas but Aker had won a solid share of the North Sea market with its bold designs for drilling rigs and production platforms. So it made some kind of sense for Siem to appear to consider purchasing the five idle boats for North Sea work. In Panama, further proceedings were set in motion to confer on Siem 'full powers for the management of the activities of Starboat'. The company was even provided with an address – albeit only a post office box, number 25078 – in Oslo. Martin Siem himself flew from Norway to France to meet the head of the CMN shipyard, seventy-nine-year-old Felix Amiot.

Amiot was sometimes called the boss of Cherbourg. When he was eighteen he opened his first factory in the Normandy seaport, and proceeded to build not ships but aircraft. He invented a new way of constructing fuselages from metal tubes and when World War One broke out his Amiot 01 fighter was in great demand by the young French Air Force. By 1918 he had two factories employing 1,400 people.

Amiot went into mass production and during the 1930's the French government ordered his Amiot 143 by the hundred. His reputation reached its zenith with the Amiot 370 which smashed eleven world speed records in 1938 and 1939. In World War Two Amiot's factories were destroyed.

In 1945, already fifty-one, Amiot changed course. He moved into timber, steel and shipbuilding, changing the name of his company from Chantiers Aeronautiques de Normandie to Constructions Mecaniques de Normandie, or CMN. He built minesweepers and coastguard vessels from wood, and trawlers and other fishing boats from steel. But Amiot's new success was based above all on his fast, powerful, and highly manoeuverable patrol boats, equipped with a formidable array of arms. Later these were given the capacity to launch missiles, and it was a version of this type, dubbed the Mivtach, which Israel had ordered.

When Martin Siem met Amiot and told him that he wanted

to buy the five remaining gunboats for Aker's North Sea work, it was not a story that Amiot can easily have swallowed. The gunboats had been designed for the Mediterranean. Although susceptible to sudden storms, the Mediterranean cannot compare with the North Sea, one of the wildest and least hospitable stretches of water in the world, as some oil operators were discovering to their cost. Siem said that the boats would be used to ferry supplies to drilling rigs and production platforms, but this, too, was scarcely credible. With their enormous engines and a top speed of more than forty knots they were ludicrously overpowered for such work, and for anyone to consider paying two million dollars each for supply boats was absurd. As one observer later commented, it was like using a Porsche to pull a cart.

Amiot can be forgiven for not voicing the suspicions he must have felt. The shipbuilding industry is a fluctuating one and in 1969 Amiot faced the very real prospect of having to lay off some of his workforce. Israel meanwhile had paid only a four million dollar deposit on the last five boats and the French government had rejected all Amiot's pleas to make up the balance of six million dollars still owed. When Siem offered to pay the full purchase price of ten million dollars for the five boats, Amiot was delighted to accept.

Even though the boats were supposedly destined for North Sea oil work, they were still classified as gunboats. And in France all arms sales have to be sanctioned by the government. It was no more difficult to win such approval than it had been to get Euratom's blessing for the fake uranium deal.

The body charged with vetting arms deals had the imposing title Commission Interministerielle pour l'Etude des Exportations de Materiels de Guerre. It was a committee – CIEEMG for short – composed of members of all the departments and ministries likely to have an interest in such sales. On 12 November Amiot visited his old friend and contact at the Ministry of Defence, General Louis Bonte. Bonte was director of International Affairs in the Ministry's arms department. He was also deputy chairman of CIEEMG, which was due to consider the gunboat sale at its meeting later that week. Bonte advised Amiot that there were two principal questions that would concern the committee: could Siem pay? And was the deal safe?

The first question was easily answered. Starboat most certainly could pay, and it would do so in cash. In the same way, Yarisal had paid cash for the *Scheersberg A* and Asmara for the uranium, thereby removing the need for any detailed examination of their backgrounds. One member of CIEEMG did look up Siem in the current edition of an international *Who's Who*. The long and impressive entry seemed to render any further inquiry superfluous.

The second question presented hardly more difficulty. Bonte said that with the publicity, not to say controversy, surrounding the Israeli gunboats, the committee would wish to satisfy itself that Israel had renounced all claim to them. A letter would suffice. In the Plumbat operation it was Asmara, one of the central parties involved, who provided the letter to assuage the anxieties of the bureaucratic mind. The most perfect irony of the Cherbourg affair is that the letter that reassured the French authorities was written by Israel herself.

On learning of the committee's requirement, Amiot got in touch with Admiral Mordecai Limon at the Israeli Embassy. Amiot explained that France sought a letter from Israel declaring that she had formally abandoned all claim or title to the five remaining gunboats. Limon solemnly obliged. At Amiot's prompting he dictated precisely the letter that the committee wished to read.

The committee duly met to consider the deal. By most accounts, its approval was a formality.

The ease with which the Paris business had been completed should have left Mossad with a feeling of contentment and an optimistic prognosis for the outcome of the scheme. But Mossad had other things on its mind. On 12 November, the very day that Amiot was conducting final negotiations with General Bonte, Mossad was preoccupied with a shattering piece of information. West German detectives had just boarded the *Scheersberg A*.

Chapter 11
Wonderful, Wonderful Copenhagen

Of the various inquiries launched into the disappearance of the Plumbat uranium, by far the most energetic was that carried out by the BKA*, West Germany's equivalent to the FBI. Even so it was almost a year before they found the *Scheersberg A*. They did so by questioning the August Bolten company in Hamburg, where they learned that Peter Koerner was the ship's engineer. The BKA doggedly tracked down Koerner's fiancee who told them when the *Scheersberg A* was next due in German waters. As the ship was en route from Tunis to Lulea, she said, it would be passing through the Kiel Canal on its way from the North Sea to the Baltic.

When the *Scheersberg A* entered the Kiel Canal on 11 November 1969, BKA agents boarded her. Their mission was to find out where the *Scheersberg A* had sailed to the previous November. None of the crew could help them, of course, and the ship's log was useless because the two vital pages were missing. Still, the agents asked Captain Cousillas a great many questions. They refused to reveal what their investigation was about, telling him only that it was 'top secret'. When they disembarked from the *Scheersberg A* they left Cousillas a puzzled and worried man.

Cousillas requested a local shipping agent to pass a message to Biscayne Traders. In it, the captain reported what the BKA had asked questions about. Cousillas said he knew nothing about the matter, nor would the BKA explain, and he added: 'Something is very wrong.'

The message to Biscayne Traders reached Mossad very

*Bundeskriminalamt.

quickly indeed, and caused much alarm. It was far too late to withdraw the *Scheersberg A* from the Cherbourg operation. She was already approaching Lulea where a special crew was being mustered. There was no time to find a replacement ship. The only possible course was to stick to the plan, and then get rid of the *Scheersberg A* immediately the operation was over.

That same day Dan Aerbel, then in Copenhagen, received an urgent message. He was told that 'his' company owned a small cargo vessel which would have to be sold as soon as Mossad gave the word. Aerbel was to put in hand immediately all the paperwork to allow the *Scheersberg A* to be disposed of at a moment's notice.

These orders posed Aerbel quite a problem. Even a Liberian company cannot dispose of its assets on the say-so of one director. Such a momentous decision would have to be ratified by other directors – and Biscayne Traders did not have any. Aerbel needed help, and fast. He called a friend in Copenhagen who he was certain could provide it.

Torben Hviid lives on the edge of the law; at times has been over it. He is a jovial, friendly man, good-looking, with sparkling blue eyes, and an engaging apparent naivete. He has a talent for devising ingenious schemes to make money. While they have usually done so, they have also brought him some hefty fines and the occasional spell in prison.

There was, for example, the '7–9–13' club that he opened in Copenhagen. Hviid said it was a social club for Danish old age pensioners, but the Danish authorities disagreed. They said it was a club for gambling, an activity strictly controlled in Denmark, and they took Hviid to court. One old age pensioner testified he had lost about £1,400 playing Hviid's slot machines. Hviid was fined the equivalent of £4,500.

His next scheme also concerned gambling. In the Tivoli gardens, the elegant pleasure park in the centre of Copenhagen, there are banks of gambling machines that use, and pay out, tokens costing twenty-five ore, about three pence. Hviid made his own tokens and sold them to the punters for a penny. The machine operators took Hviid to court and were amazed to discover there was nothing in Danish law to stop him. Hviid offered a sporting compromise: he said he would buy his own tokens back from the operators for a halfpenny each. The operators refused, preferring to dump them by the sack load into the Baltic Sea. In all, Hviid sold fifteen million tokens – and drove a Rolls Royce around Copenhagen on the profits – before the operators installed photoelectric cells to distinguish the fake tokens from their own.

Hviid's next venture was less amusing. He dreamt up a loan sharking scheme which cost some of his customers a true interest rate of 400 per cent. Hviid was prosecuted and sent to prison for a year.

In the late 1960's Hviid's business was sex. In 1967 Denmark legalized pornography and he joined the rush to publish hardcore magazines. Denmark was soon saturated with them and Hviid looked for export markets. The nearest was Germany, the only catch being that pornography was still illegal there. It proved so easy to smuggle magazines into Germany that Hviid's liking for publicity got the better of him. He allowed a Danish newspaper to photograph him as he drove across the border in a car full of magazines and was arrested by the German police on his very next trip. After a brief stay in Hamburg jail, he made another smuggling attempt and was arrested again. This time he was in prison for four months.

Aerbel had known Hviid since they were at school together in Denmark. They also had both sold furniture for the same firm in 1961. Aerbel enjoyed Hviid's somewhat roguish company, and usually looked him up whenever he was in Copenhagen. On the other hand, Hviid's criminal record, and his bravado, made him a totally unsuitable candidate for office in Biscayne Traders. His involvement could excite police interest in the company, and in Aerbel, which was the last thing Mossad wanted. Yet Aerbel did not hesitate to approach Hviid. On 12 November Hviid assumed the mantle of company secretary to Biscayne Traders Shipping Corporation. That afternoon he went before the notary public of Copenhagen and certified that there had been a special meeting of the company's board of directors and that Aerbel

– using the name 'Dan Ert' – was entitled to 'sign all papers concerning the sale of the vessel *Scheersberg A*.'

Having helped Aerbel with his little legal problem, Hviid now felt it reasonable to ask Aerbel for a reciprocal favour. Chastened by his two spells in a German prison, Hviid had decided that printed pornography was a mug's game. He turned to the real thing. He had already opened one club, the 'Intime', which staged live sex shows. Now he wanted to start a second club, named 'Love In', to meet the needs of people in search of sexual relief. Beneath his club for old age pensioners he had separate premises available which had already been equipped with booths where hostesses could take clients for a massage, or more. Only a few legal formalities remained before the club could open. Aerbel agreed to help.

So, just three days after Hviid became the nominal secretary of Mossad's shipping company, Mossad's eager agent served as the legal witness that allowed 'Love In' to go into business. Aerbel seems to have got the better of the deals. In return for his services he was allowed into the club for free and, in Hviid's words, he became a 'very regular customer.'

Two days after these legal proceedings had been completed in Copenhagen, the *Scheersberg A* reached the Swedish port of Lulea. It is a picturesque town, set among birch forests and shimmering inlets on the coast of Swedish Lappland, with the Finnish border only fifty miles away. But Captain Cousillas gave little thought to the scenery. Having made this marathon journey at the behest of the *Scheersberg A*'s new owner, he had now received the rather disconcerting news that the entire ship's complement – captain, officers and crew – were to be laid off.

Cousillas of course had not been on the *Scheersberg A* in Rotterdam a year before, the last time this had occurred. Chief Engineer Peter Koerner had. There was much that was strange about that previous episode, Koerner began to reflect, and much that was puzzling now. The new crew had started to arrive in Lulea and Koerner noticed that they were mostly Icelanders and Danes. Koerner wondered why anyone should go to the trouble and cost of bringing them to Lulea – especially as Yarisal or the new owners would be landed with some hefty travelling expenses for the old crew to get home.

Captain Cousillas could add little by way of explanation. He and Koerner went into town together to pick up their tickets for the next plane to Stockholm, the first leg of their respective journeys home. That night Cousillas took Koerner and the ship's three other officers for a farewell meal in the large and stately dining room of Lulea's comfortingly traditional Stadshotell. Cousillas told his guests that the new owner of the ship was at least picking up the tab.

On 22 November the *Scheersberg A*, which had taken on a new captain and crew, but which had not been renamed the *Sofa* or anything else, sailed out of Lulea and headed south.

Chapter 12
Home to Haifa

In November 1969 Cherbourg was wearing an increasingly curious aspect. It is a compact town, huddled around the harbour and dockyards that are its main reason for existence, and it was hard to overlook the presence in the community of an ever-growing number of Israelis.

Some had been there since 1966 when construction of the Israeli boats had begun. They were mostly officials and technicians. Many had brought their families to Cherbourg and some sent their children to French schools. But as the last gunboats neared completion – the eleventh was launched on 14 October, the twelfth was due in the third week of December – more and more sailors arrived. CMN installed some of the officers at the red-bricked Hotel Atlantique, no longer a hotel but the company's headquarters and main design office, close to the waterfront. Most of the seamen were accommodated in boarding houses and private lodgings throughout the town.

This perplexing situation was the result of a compromise that had been hammered out after the Israeli cabinet had authorized Mossad to take charge of fetching the gunboats. The Navy conceded the operational design of the mission to Mossad, but insisted that it was going to sail the gunboats home itself.

Mossad had been compelled to devise a story to explain why there were so many Israeli seamen in Cherbourg when the gunboats had supposedly been bought by a Norwegian Company for work in the North Sea. The official reply to anyone who asked was that Starboat had requested the Israeli Navy to deliver the boats to Norway on its behalf. It was a farcical explanation but no one in Cherbourg, least of all the local press, was inclined to pursue the matter.

The British press displayed no such inhibitions – and in December came close to wrecking the deal. A party of journalists from England was invited to Cherbourg on a junket to celebrate the launching of France's latest nuclear submarine, the *Terrible*. They were puzzled to come across so many Israelis, and Anthony Mann, Paris correspondent of the London *Daily Telegraph*, filed a story about the mystery on 16 December – the day the twelfth and last boat was launched.

Mann wrote of the 'extraordinary situation' in Cherbourg, with 'five Israeli crews of thirty-three men each' waiting there. Mann said that the future of the five gunboats seemed 'obscure' but added that an official of the Israeli Embassy in Paris had told him: 'We know of no plans to sell these ships to anyone else'. Mann's most telling line came at the end. 'It is understood that some Israeli naval personnel will return home on Christmas Day'.

In Paris, Mann's article caused consternation. France's external intelligence service, the SDECE*, clipped the report and sent it with a sharp memo to the Ministry of Defence, asking what was going on. Once again a sympathizer in the Ministry of Defense tipped Israel off. And when Admiral Limon learned of the danger to Israel's carefully laid plans he sent word to Mossad: it was time for a little less stealth and a little more daring. The original scheme had called for the gunboats to sail at least some of the way towards Norway in order to preserve the story that they were destined for Siem. The time for such niceties was passed, Limon said. When the boats left Cherbourg they should – to put it crudely – turn left for Israel instead of right for Norway.

Early on 24 December Admiral Limon left Paris in his official black Jaguar, its diplomatic licence plate – 59 CD 59 – openly displayed. It was a show of flamboyance that seemed the best antidote to the anxiety that even Limon was beginning to feel: Had Israel left it too late?

Towards lunchtime the Jaguar pulled up at the Cherbourg

*Service de Documentation Exterieure et de Contre-Espionnage.

Sofitel, a hotel in a barren slab of a building at the heart of the harbour area. Limon took one room in his own name and one for his chauffeur, who gave his name as Victor Zipstein. The hotel asked Zipstein if he and his boss would be wanting the Christmas Eve dinner, a traditional French celebration. Zipstein said they would not. 'Jews don't celebrate Christmas,' he explained. 'We're here on business.' Soon afterwards Limon departed to have lunch with Felix Amiot.

After lunch Zipstein drove Limon to the waterfront. Limon walked the short distance across the quays to where the gunboats were tied up in a neat row of five. There he met the man who would take charge of the fleet on its long journey home, a handsome and imposing Israeli navy officer who used the name Commander Ezra. Limon told him it was imperative for the five boats to leave as soon as darkness fell.

Ezra replied that this would be very difficult, and pointed out the foaming white caps on the sea beyond the harbour walls. The waves were running at between fifteen and twenty feet high; the gunboats had after all been built for the Mediterranean, not the English Channel or the Atlantic in winter. Limon could not conceal his impatience. When was the next weather forecast due?

The British brought Israel good news. The Southampton meteorological office predicted later that afternoon that the wind would ease. Ezra agreed to leave.

The lifting of the five gunboats had been coolly and meticulously planned, with moves made in places as far apart as Tunis and Panama, in a game that had gone on for months. That evening the final preparations for departure bordered on panic. The order to round up the five crews was hastily spread through Cherbourg. Dinner bookings went uncancelled and shaving tackle was left in washbasins as Israelis hurried to the quayside in ones and twos.

Until that point the fiction that the boats were bound for Norway was still being maintained. All bureaucratic requirements had been fulfilled. An export permit for the boats had arrived from Paris and was safely lodged in the office of the Cherbourg dockyard agent. But there were some formalities left to complete. Cherbourg's harbour regulations include a rule that all vessels must give twenty-four hours notice of their intention to leave. Limon decided to ignore it. If their cover really was about to blow it would be far too risky to alert the French to what was going on.

The route out of Cherbourg also presented considerable hazards. The five boats were tied up inside a harbour that was formed by the jaws of two jetties. Beyond that was Cherbourg's outer harbour, guarded by a long sea wall which Napoleon had built. Beyond that lay the open sea.

The problem facing Limon and Ezra was this. The western jetty that helped form the inner harbour was also part of the perimeter wall of Cherbourg's naval base. As the home of France's nuclear submarines, the naval base was heavily guarded, even in the early hours of a Christmas morning. The customary route out of Cherbourg lay past the naval base and through a gap at the western end of Napoleon's sea wall. If the gunboats chose that route they could scarcely avoid being seen.

There was only one alternative. That was to veer *east* after leaving the inner harbour and aim for a narrow gap at the other end of the outer sea wall. No one should see the boats if they left that way. But there were other dangers. The channel was dredged to a depth of only ten feet, giving the gunboats clearance of less than two feet, and the narrow gap in Napoleon's wall was made more perilous by the submerged rocks of the nearby Ile Pelee (the 'bare island'). The route was bleakly dismissed on the official harbour charts as 'interdit' – forbidden. But that was the route, Limon and Ezra decided, the five boats would have to take.

Boldness brought Israel her reward. A few respectable burghers of Cherbourg on their way to midnight mass did see the last Israeli sailors, clutching packets of American cigarettes, hurrying to the quayside. The only other witness was Napoleon himself: or rather his statue, astride a bronze horse, pointing out to sea from a vantage point above the harbour. The plinth of the statue was inscribed: 'I have resolved to repeat at Cherbourg the marvels of Egypt'. Otherwise, Cherbourg simply awoke to find that the gunboats, so long a part of the landscape, had disappeared.

The *Scheersberg A* had spent most of December in the Mediterranean waiting to play her part in the operation. From Lulea she had sailed to Lisbon, arriving on 2 December. She left three days later and called next at the French Mediterranean port of Toulon. She stayed there for just twenty-four hours before putting to sea again. She now spent nine days lazing her way down the Spanish coast to the port of Almeria, 150 miles from Gibraltar.

Ever since the Cherbourg plan had been drawn up, Mossad had intended to use the *Scheersberg A* to refuel the gunboats. Exactly where that would take place had been left undecided. It would depend on how far the boats went towards Norway before turning back. It could perhaps have been carried out somewhere to the north of Gibraltar, the first port used by boats one through seven as they made their legitimate way to Haifa. But, with the cover story in growing disarray, everything changed.

The *Scheersberg A* had already arranged to put in at Almeria on 20 December to collect a load that Mossad had organized and to await further orders. On 21 December the orders came. The *Scheersberg A* was to proceed towards France at all possible speed. She must reach the Bay of Biscay by Christmas Day.

Loading was hurriedly completed and the ship left Almeria the next morning. She declared her destination as Brake, a port close to the mouth of the River Weser in West Germany. She told the Almeria harbour authorities that the drums she had taken on board contained industrial sand. In fact, they were full of diesel fuel.

Sailing close to her maximum of 12½ knots, the *Scheersberg A* passed through the Strait of Gibraltar that evening. The next day, 23 December, she reached the southern tip of Portugal, then headed up Portugal's west coast. She arrived at her destination off the northwest tip of Spain in the early hours of Christmas morning. The nearest port was La Corunna. To the east were the wild waters of the Bay of Biscay. Somewhere to the south lay Cape Finisterre, which means the end of the earth. In this forbidding place the *Scheersberg A* hove to.

The meeting point had been carefully chosen. With the Norwegian fairy story near collapse, it was no longer possible to assume that the five gunboats would be able to leave Cherbourg with full tanks. From Cherbourg to this position via the Bay of Biscay was some 500 nautical miles, one-third the gunboats' range at their customary cruising speed of twenty-two to twenty-three knots. And from here to Haifa was 2,600 nautical miles. That was the longest distance the gunboats could be certain of covering with only one more refuelling.

Because of the bad weather in the English Channel which delayed the gunboats' departure, the *Scheersberg A* had twenty-four hours to wait. Shortly before dawn on 26 December the gunboats arrived, being guided over the final stretch to their rendezvous by walkie-talkie. The oil contained in the drums carried by the *Scheersberg A* weighed 150 tons, three-quarters the Plumbat load, but the transfer was far more awkward in the Atlantic than it had been in the eastern Mediterranean a year earlier. It was almost evening before the task was completed. Then the *Scheersberg A* and the gunboats parted. The gunboats continued south for the Strait of Gibraltar and the Mediterranean, while the *Scheersberg A* headed north for her declared destination of Brake.

The refuelling operation, despite its obvious hazards, had been conducted with one enormous advantage: the French government did not yet know the gunboats had gone. Its blissful ignorance was the result of an extraordinary conspiracy of silence on the part of the town of Cherbourg.

Cherbourg is served by two newspapers: *La Presse de la Manche* and *Ouest-France*. Reporters on these publications also act as correspondents – known in the trade as stringers – for other French newspapers and agencies. The journalists of Cherbourg, reversing their normal role as disseminators of information, now suppressed the news about the flight of the gunboats.

Marc Gustiniani, managing editor of *La Presse de la Manche*, unashamedly justified his newspaper's decision soon afterwards: 'As the result of a tacit agreement with the manufacturer, the representatives of the press kept silent. We were guided by our unique concern not to harm the activity of the most important local industry, which employs more than 1,200 people. We would have observed exactly the same position if the shipyard had been working for any other country.'

Forty-eight hours passed before the news broke. Somebody in Cherbourg tipped off the Paris office of Associated Press. Within minutes the story was on the wires and the Paris newsdesks hastily obtained confirmation from their hitherto mute Cherbourg colleagues. Then they called the Ministry of Defence. The Ministry was puzzled at the fuss and at 2:12 am on 27 December it issued a statement. The boats were 'the subject of a normal commercial arrangement with a Norwegian company.' And so far as the Ministry knew, Norway was where the boats had gone.

The Norwegian embassy soon exposed this for the nonsense it was. Later that day it issued a statement. Starboat was not a Norwegian company. The boats had no right to fly the Norwegian flag. And they were not expected in Norway. Any remaining doubts where the boats were bound were dispelled that evening when Lloyd's agent in Gibraltar reported seeing the five boats race through the Strait into the Mediterranean. His news was broadcast on French radio that night.

France had been humiliated. But as Pompidou himself told his aides, unknowingly echoing Mossad's own judgment, he was not de Gaulle. There would be no grand military gesture in the General's style, no attempt to arrest the gunboats on the high seas. The only planes which swooped on the gunboats as they hurried east were those of the world's press vying to take pictures from the closest range. Later Pompidou, raging at the 'unbelievable carelessness and intellectual complicity' of his officials, fired two senior members of CIEEMG, the committee which had approved the deal, and demanded Admiral Limon's banishment to Israel. It all came too late to remove the gloss from Israel's victory.

Under cover of darkness the gunboats refuelled for a second time off Sicily. On 30 December, the *Scheersberg A* slipped quietly into Brake. Late in the afternoon of 31 December the five gunboats made their triumphant entry into Haifa. There was no longer any point in secrecy and Israel turned the occasion into a public celebration, one more episode to be written in Israel's short but heroic history.

General Moshe Dayan, surrounded by reporters and cameramen, was at the harbour to welcome the boats home. The news of their arrival was flashed to Prime Minister Golda Meir at a meeting of the Jewish Agency in Jerusalem.

Such was the excitement that Israel took the unusual step of permitting Commander Ezra, the hero of the hour, to answer questions at an impromptu and chaotic press conference. The commander did his best to stick to matters of seamanship. The waves in the Bay of Biscay had been 'eighteen to twenty feet high', but then 'December is not the best month to travel at sea'. He had been on duty for six days, snatching no more than ten minutes sleep at a time.

More awkward questions intruded: who were the ships for? Commander Ezra understood they were for Netivei Neft, Israel's oil exploration corporation. Indeed, Netivei Neft had sent a representative to the quayside but as he insisted on speaking Hebrew, foreign reporters turned back to Commander Ezra. What would the boats be used for?

'I don't know and I don't care. I'm a naval officer. I can't tell you if these ships are suitable for oil prospecting.'

Ezra was asked if the five boats had been refuelled during their journey. 'Yes,' he replied tersely. 'I refuelled.' Pressed for details, he said that the refuelling had been carried out by a ship that was 'not a tanker'. Pressed again, he announced: 'I am tired. I think I am going home to bed.'

Commander Ezra's reticence on this point was understandable. Israel was happy to bathe in the inspiring publicity of the gunboats' dash for Haifa. But as Ezra had been warned, the question of the refuelling was the one above all that Israel did not wish to see explored. If the part played by the *Scheersberg A* came to light there would certainly be further inquiries into the ship's background. And the *Scheersberg A* was the strongest single link between the missing 200 tons of uranium and the state of Israel. It was time to get rid of the evidence.

Chapter 13
Loose Ends

It is clear that Mossad decided to divest itself of the *Scheersberg A* as rapidly as possible after the Cherbourg operation. It is also clear that Dan Aerbel did not have sufficient experience in the shipping business to carry out this task by himself. That, at any event, seems the best explanation for the reappearance in the story of the elusive Burham Yarisal.

In the days between Christmas and New Year's Eve 1969 the Hamburg shipbroker Uwe Moeller received yet another telephone call from Yarisal. Moeller had not heard from Yarisal since selling him the *Scheersberg A* in 1968. Yarisal said that he had since sold the ship, and the new owner now wanted to do the same. Yarisal asked Moeller to find a buyer, stressing that it was a matter of some urgency.

The *Scheersberg A* was on offer at a bargain price, almost £25,000 less than Yarisal had paid for it, and Moeller's firm found a potential buyer very quickly indeed, a Greek ship owner named Georges Franjistas. Unfortunately, Aerbel's preparations to sell the ship had not been thorough enough. Franjistas wanted better evidence that Aerbel was entitled to dispose of Biscayne Traders' sole asset. On 5 January 1970, Aerbel visited the Copenhagen notary public, armed with the papers that had been signed in October and November, together with a new document that summarized what they contained. Aerbel signed the summary and Mr Peytz, the notary, added his signature and his large and impressive seal. Not even this sufficed. Aerbel had to go on to the Greek Consul in Copenhagen who certified that Mr Peytz's signature on the document was genuine and affixed his own seal, rather smaller than Peytz's – impressive nonetheless.

While all this was going on Uwe Moeller took the opportunity to spend a week in Norway, tying up another deal. Late on 16 January he received a telex at the Caledonian Hotel in Kristiansand. It was from his firm and it told him to go to London at once to meet the new owner of the *Scheersberg A*. The command put Moeller in some difficulty. It was Friday night, he was expected home for the weekend and he was carrying very little cash. Now he would have to leave at six the next morning to fly to Copenhagen in time for a connecting flight to London. It took a lot of late night telephone calls to arrange for money and a ticket, and he arrived in London at lunchtime the next day in a rather weary state.

There was no time for recuperation. As soon as he walked inside the airport terminal he was paged and asked to go to the information desk. Dan Aerbel was waiting for him. He introduced himself to Moeller as Dan Ert and said that he needed his help in selling the *Scheersberg A* because it was his first ship. That afternoon the two men took an Iberia Air Lines flight direct to Bilbao, a port on the Basque coast of northeast Spain.

The *Scheersberg A* was already in Bilbao, having arrived two days earlier. Moeller and Aerbel checked into the Hotel Carlton and early the next morning walked down to the harbour to find she was in dry dock being surveyed for Georges Franjistas, due in Bilbao soon. There was little wrong with the ship. She had a gash on her bows which required minor repairs and a dash of paint, and she needed a new anchor chain. But Bilbao's reputation for efficiency was not high and Moeller and Aerbel spent all of the next two days at the harbour supervizing the repair work. Their only relaxation was a visit to a night club where Aerbel talked a little about himself. He said he owned a furniture shop in Rome and that he was compelled to sell the *Scheersberg A* because the furniture business was taking up most of his time. He said he might buy another ship, one day.

The dry dock work was completed by 20 January. Georges Franjistas arrived and went on board. He looked the ship over, found everything in order and then instructed the First National City Bank in Hamburg to pay Biscayne Traders the purchase price of $235,000. He transferred the *Scheersberg A* from Liberian to Greek registry and renamed her *Haroula,* after one of his daughters. The *Haroula* hoisted the Greek flag and sailed from Bilbao on 27 January.

There remained one last Plumbat clue to dispose of: Biscayne Traders Shipping Corporation. Aerbel could simply have ignored the problem. Under Liberian law any company whose fees and taxes are unpaid for two years is automatically dissolved. That, however, was not Aerbel's style. A meticulous man, he felt it necessary that the company be closed down with all due formality.

Aerbel was still seeing a lot of Torben Hviid. From time to time more official signatures were required in the large book that served as the register of 'Love In.' Three times in the summer of 1970, on 9 June, 14 July and 25 August, Aerbel signed the register as the club secretary. So, when it came to closing down Biscayne Traders it was natural for Aerbel to turn to Hviid again. He asked his friend if he knew a lawyer who could help with the formalities. Hviid recommended Kaj Lund, a lawyer who worked from a very respectable book-lined office close to Copenhagen's Royal Palace.

Lund had helped to set up several of Hviid's enterprises, including 'Love In.' But he had no idea how to close down a Liberian company and Aerbel had to enlist the assistance of the Liberian Services office in Zurich. On 17 November Liberian Services sent Lund a letter which began: 'Mr Dan Ert, President of Biscayne Traders Shipping Corporation has asked us to write to you...' The letter detailed the procedures Lund should follow. They covered almost two pages. Liberian companies are a great deal harder to dissolve than to set up.

In due course, on 8 January 1971, a 'special meeting of the board of directors' of Biscayne Traders was held. The minutes solemnly record that three people were present: Aerbel who took the chair, Lund who acted as secretary, and Hviid, now promoted to company treasurer. The minutes continue: 'It was decided by all present at the meeting that Biscayne Traders Shipping Corporation should be dissolved.' They particularly noted that as 'holder of all shares of the corporation' Aerbel gave his consent.

There was still some way to go. The next step was to draw up a Certificate of Dissolution. On 18 May lawyer Lund escorted Aerbel and Hviid to the office of the Copenhagen notary public to witness their signatures. The notary checked Aerbel's passport – the one he carried in the name of Dan Ert – and signed the dissolution document himself. Lund sent it to Liberian Services who promptly sent it back, telling him he must authenticate the signature of the notary. Lund went to Harald Nissen, Liberian Consul General to Denmark, who duly certified that the notary's signature was genuine. Lund sent the papers back to Liberian Services. Finally, on 24 September 1971, one J. Rudolph Grimes, Secretary of State for the Republic of Liberia, added his signature to the Certificate of Dissolution. Biscayne Traders Shipping Corporation had ceased to be.

So, too, had 'Love In'. The Copenhagen police had visited the club several times, as had the Danish tax inspectors. Hviid decided to close the club and he called on Aerbel to sign the final entry in the official register. The move was too late to save Hviid from prosecution. He eventually appeared in the Copenhagen city court to face a spectacular range of charges: pimping, procuring and evading sales tax. He was cleared on the first two counts but convicted on the third and fined £3,300.

At the end of 1971 Dan Aerbel left Europe and went home to Israel.

Now that it no longer mattered, Mossad told him something of the role the *Scheersberg A* had played in the Plumbat and Cherbourg operations. It did not tell him everything – just enough, as is customary, to persuade him that his work had been important and appreciated. Such encouragement, naturally, made him want to continue working for Mossad.

Mossad's gratitude seemed well-placed. Aerbel had followed his instructions to the letter in disposing of both the ship and Biscayne Traders. The irony was that he had done his job too well. Aerbel would have done better to allow the company simply to fade away, for in following the cumbersome rigmarole of dissolution he scattered clues to the truth about Plumbat.

That would not have mattered if Mossad had continued to employ Aerbel on similar assignments. But two years later Mossad sent him on a mission that was totally beyond his capabilities. The consequences for Aerbel and Mossad were disastrous.

Chapter 14
The Salesman Was a Spy

Dan Aerbel should never have been selected as a secret agent in the first place. It is true that he was well-motivated, identifying firmly with Israel's cause. As a personable and widely travelled young man he had a certain talent for collecting scraps of useful information. But from the very beginning he displayed weaknesses which made him unsuitable for a career in the business of espionage. He was indiscreet, and had an undue liking for melodrama. Under stress these flaws in Aerbel's personality became fatal weaknesses. When it came to the crunch in Norway in July 1973 his instinct for self-preservation overrode all other considerations.

A clue to Aerbel's character is to be found in the variety of names he used during his ten years with Mossad. There was the name he was born with, Erteschick; the versions he shortened that to, first Ertz, then Ert, also Ertl; the name Aerbel which he came to use in Israel; and the rather selective Arbel which he had inscribed on visiting cards for the occasions when he wanted to conduct business in English. One explanation is that these switches were an elaborate and carefully planned part of his Mossad cover. But they occur too frequently and too randomly for that. He seems to have switched his name as frequently as he switched his role until in the end he became all things to all men. The question left unanswered is whether Dan Aerbel ever discovered where his true identity lay.

Dan Erteschick was born in a suburb of Copenhagen, capital of Denmark, on 28 February 1937. His father Mario was a prosperous businessman who owned a clothing factory. Dan, the middle of three children, was brought up in a spacious white stone house on the sweeping shore of the Baltic some ten miles north of Copenhagen. A short way beyond was Helsingor, the Elsinore of Shakespeare's Hamlet. The name of Dan's mother was Gertrud.

The Erteschicks were respected members of Denmark's Jewish community, and Dan's would have been a privileged childhood had World War Two not intervened. The Nazis invaded Denmark in April 1940 and at first hoped their occupation would be received peacefully. But opposition from the Danes mounted and by 1943 there were frequent riots and acts of sabotage. In August the Nazis declared martial law and the German army took control. In October 1943 the roundup of Jews began.

For the Erteschicks there was an obvious route to safety close at hand – the ferry from Helsingor to neutral Sweden, just two miles away across the Baltic. But the Nazis realized it too, and the Erteschicks went into hiding as they awaited their chance. For six-year-old Dan it was an experience of utter terror. He and other Jewish children were hidden in a school cellar, and to increase their chances of avoiding detection they were bricked in. Dan spent two months in fear and semi-darkness before the family could escape to Sweden. Those two months left Dan with two deeply-scored psychological marks: a total identification with the plight of the Jews; and a fear of being confined, an acute claustrophobia, that would never leave him.

After the Nazis' defeat the family returned to Denmark and Mario Erteschick built his business again with a chain of fashionable furniture stores which he named Diva. At school Dan was considered bright and industrious. He showed some interest in economics and foreign languages and went on to study at college in the United States. He left before graduating and returned to Denmark.

On his return he found that his father was coming to an important decision. Mario Erteschick was eighty-six and felt that it was time to go to his spiritual home, Israel. It was a convenient moment to leave, for the Danish government had embarked on a somewhat drastic policy of environmental improvement on the coast north of Copenhagen. In order to create clear views across the Baltic it was buying up houses

The *Scheersberg A*: Now she has a different name but her old identity is barely hidden. In December 1968, in halcyon weather, she headed for the waters between Cyprus and Turkey – and for a silent rendezvous with an Israeli freighter.

Bryan Wharton/Sunday Times

DAN AERBEL (right) joined Mossad in 1963. Nine years later he was sent on a mission totally beyond his capabilities – and he was caught up in murder (inset right). The consequences for Mossad, and its director ZWI ZAMIR (far right), were disastrous. But until then Aerbel was a successful agent and he played a crucial role in Operation Plumbat. Circumstantial evidence suggests Aerbel spotted HERBERT SCHULZEN (below) who became one of Israel's very special friends. Schulzen's company, Asmara Chemie, bought 200 tons of uranium and won permission to ship it from Belgium to Italy. The uranium was consigned to a paint company run by FRANCESCO SERTORIO (below centre) but it never arrived. Investigators could find no trace of the missing cargo: one of them, armed with a Geiger counter, went to Asmara's premises near Wiesbaden (below right) only to discover the building was completely empty.

MYR

Dagbladet, Oslo

Ali Hassan Salameh (above left) was Israel's implacable enemy and the 'hit team's' prime target. Mossad tracked him to Lillehammer where Ahmed Bouchiki (above right) met a stranger; the meeting cost Bouchiki his life. Sylvia Raphael (right) was one of the Israeli agents arrested for murder but she stuck to her cover story. It was Dan Aerbel who cracked: there was a long silence, then he told an astonished Inspector Steinar Ravlo (below) every secret he knew.

Tor Gulliksrud/Dagbladet

that stood in the way and the Erteschick home was one of them. Mario sold the white stone house for 350,000 kroner (£18,000) and got a good price for his chain of furniture stores, too. In 1962 Mario and Gertrud Erteschick emigrated to Israel. Their son, then twenty-five, stayed behind. Soon afterwards he was recruited by Mossad.

At that time, Mossad was renewing its European network. Its helpers and allies from the days of the fight against the Nazis were growing old and soon would need replacing. In its search for recruits, Mossad asked for recommendations from the leaders of Jewish communities, and almost certainly heard about Dan Aerbel in this way. He seemed ideal agent material. He was from an old and established family that had shared in the sufferings of the Jews during the war. He was already well-travelled and fluent in several languages. His parents had just come to Israel but he had decided to remain in Denmark to start a career of his own. Above all, he was well-motivated. He was loyal to the concept of Israel. He also had told several of his friends that he hated Arabs.

The evidence suggests that Mossad's first tentative approach was made in the autumn of 1962. At around that time Aerbel had answered an advertisement in a Copenhagen paper for the post of furniture salesman. It had been placed by a young, expanding company named Danskform, which owned a factory in northern Denmark and had just opened a smart office in the Town Hall square in the centre of Copenhagen. Aerbel went for an interview and made a good impression. 'I was very pleased to get someone so eager,' says Aldo Eggers-Lura, one of Danskform's directors. Aerbel got a job which paid a monthly retainer, plus commission.

So it was with some surprise that Eggers-Lura learned from Aerbel soon after he joined the firm that he intended to visit England early in 1963. 'I was a bit peeved,' Eggers-Lura says, 'because we had no business in England at all.' Aerbel said he was obliged to go to London to visit a relative. Because Aerbel had already proved himself an enthusiastic salesman, Eggers-Lura raised no further objection.

Aerbel was away for six weeks. He sent a rather enigmatic card to the girls in the Copenhagen office which said he was 'riding around London in a Rolls Royce.' Back in Demark he told a fellow salesman that he had stayed in a luxurious flat put at his disposal by a wealthy Jewish businessman. When Eggers-Lura reminded Aerbel that he had said he was going to London to visit a relative, he replied: 'Well, it was something like that.'

It was almost certainly something quite different. Having told Jewish community leaders in Denmark that he was willing to perform services for Israel, Aerbel was sent to London to meet the 'wealthy businessman.' The businessman entertained Aerbel lavishly, and he was duly flattered. At the same time his attitudes and capabilities were more closely scrutinized. Given preliminary approval as a suitable Mossad recruit, he was asked to see what he could find out for Israel in his travels in Europe. It all seemed straightforward to Aerbel, and he happily agreed.

But Aerbel was already displaying the fatal flaw that was to prove his – and Mossad's – undoing. Aerbel liked to brag. That was the reason for the post card mentioning the Rolls Royce, and the boast about the luxury flat. He also dropped other hints about his visit to London being for some 'higher purpose'.

Aerbel made up for his temporary absence from Danskform by working with renewed vigour. Scandinavian furniture, with its clean designs in oak and pine, had suddenly become popular in Europe, especially among the United States forces serving there. For Danskform, like other European companies, America's military bases were a highly desirable market. It showed its furniture at a series of exhibitions, arranged through the ubiquitous PX's, at US bases in Spain, Italy, France and – especially – West Germany. Danskform sold the highest volume of its furniture in and around the European headquarters of the US Air Force at Wiesbaden.

Business was good but the selling procedure was cumbersome. When an exhibition was over the salesman would have to wait until the military could arrange transport to the next base. The proposal for an improvement came from Dan Aerbel: why didn't Danskform set up a travelling exhibition of its own? Eggers-Lura soon found a bargain in a single-decker bus that he bought from a kitchen-equipment firm for £1,800. Danskform's sales staff thought the bus a great improvement. It saved them considerable waiting time and gave them greater freedom of movement. Indeed, Eggers-Lura was occasionally annoyed to find that Aerbel deviated from the firm's carefully prepared touring schedule. But once again he was prepared to forgive Aerbel. The work of a travelling salesman was hard, and Eggers-Lura concedes that he did not pay his salesmen too well: 'He was the only one who didn't complain about the work or the wages.'

One reason why Dan Aerbel was so enthusiastic in his work is suggested by fellow salesman Torben Hviid, the latter-day pornographer and sex show impresario. He had been at school with Aerbel and started working for Danskform at roughly the same time. Hviid and Aerbel sometimes went on selling trips together and Hviid remembers clearly where Aerbel's interests lay: 'He went after the girls, and he wanted a good time.' Aerbel was good-looking in a sensual way. He was tall and well-built, if slightly overweight. He had fleshy lips, rather full cheeks and a good head of curly black hair. He had a bold stare and was not afraid to meet people full in the eyes. When he wanted to, Hviid says, he could be very charming indeed.

But there was one other strong reason why the life of a salesman suited Aerbel. He was eager to prove his use to Israel and in his travels was acquiring valuable information for Mossad. Armed with a pass issued by the United States authorities at Wiesbaden, Aerbel had access to almost any American base in Europe. Security varied from base to base. Some had three checkpoints to pass through, and the guards would conduct body searches and ask questions like: 'Have you dyed your hair?' But once inside, an intelligent man who kept his eyes and ears open could learn quite a lot. After the selling session Aerbel would be invited to the mess for a drink. Young officers away from home found the pleasant young Dane, with his fluent English, a good listener.

An indication of the type of information available to Aerbel comes from another of his fellow salesmen, a Dane named

Per Bonnesen. With his attractive Australian wife Ann, he also toured American bases for Danskform. Once the Bonnesens visited a base in France soon after General de Gaulle had told the United States that planes armed with nuclear weapons could no longer fly from NATO bases on French territory. The United States had apparently agreed. 'Don't take any notice of the General,' one American pilot told the Bonnesens. 'We're flying them all right.' Torben Hviid says that such indiscretions were common: 'In Germany and France a lot of the PX officers were Jewish, and they talked politics all the time.'

Besides picking up informed gossip, Aerbel was also in a position to spot potential 'friends' – like Herbert Schulzen of Asmara Chemie – who might be useful to Israel. Events suggest that Aerbel performed this role well, for in the summer of 1964 his Mossad apprenticeship seems to have come to an end. Aerbel announced to a rather disappointed Eggers-Lura that he was leaving Danskform – in order to go to Israel. This news surprised Aerbel's friends. Only two years earlier, at the time of his parents' emigration, he had explained his decision to stay in Denmark by saying that he was afraid Israel would be invaded by her Arab neighbours. He did not now explain his change of heart.

It may have been that Mossad, impressed with Aerbel's efforts so far, wanted to take a closer look at him. In any event, Aerbel moved to Tel Aviv where he soon found a new business partner, a South African named Geoffrey Tollmann. Together they opened a shop in Tel Aviv's fashionable Trumpeldor Street. It was named Danish Interior and it sold Scandinavian furniture.

The partnership was not a total success. Tollmann found Aerbel a restless character: 'He wasn't working very efficiently. I thought he was some kind of dreamer.' And little more than a year after the business had been formed, Aerbel told Tollmann that he was pulling out. 'He told me that he wanted to travel before settling down to have a family,' Tollmann says. Aerbel sold his shares in the partnership and made a decent profit on his original stake. 'We weren't sorry when he went,' says Tollmann. 'He never really seemed to fit in.'

Aerbel moved to Paris. He took with him his new wife Debra, the Israeli-born daughter of an American professor. They rented an apartment in the university quarter of the Left Bank and Debra started a course at the Sorbonne.

The choice of Paris was no accident. Mossad had selected the city as its European headquarters, for a convincing array of reasons. In the early 1960's France's relations with Israel were at their best, and Mossad was given tacit hints that its presence would be tolerated. A number of Arab organizations had also chosen Paris as their headquarters, which made it a convenient place for Mossad to find out about their activities. Mossad's agents liked Paris for solid human reasons, too: it does contain the best restaurants in the world.

In Paris Mossad established itself as a virtual second embassy. It secretly negotiated with the embassies of governments with whom Israel had no formal ties, such as Turkey, Iran, Jordan, and Morocco. It liaised with the European companies supplying Israel's strategic needs, not least among them a French front company which helped to build and equip Dimona.

At the same time Mossad built the apparatus it needed for its more nefarious purposes. It rented apartments, offices, and houses: at 50 Rue Remy Dumoncel in the 14th *arrondissement*, for example, it built a somewhat sinister totally soundproof room. In another apartment it stored a full range of up-to-date electronic surveillance equipment. In Apartment 3, at 5 Quai Louis Bleriot, in the 16th, it installed one of its most successful operatives: an attractive thirty-year-old South African named Sylvia Raphael, who posed, very convincingly, as a Canadian freelance photographer under the alias Patricia Roxburgh. As such she had carte blanche to travel the world, and her collection of photographs in Mossad's files grew steadily. Another agent lived at 124 Avenue Wagram in the 17th. A British Jew, born in Manchester as Jonathan Isaac Englesberg, he had since become the businessman Jonathan Ingleby, with a genuine British passport – number 832427 – in that name, and a reputation as a successful specialist in trade between France and Eastern Europe.

It was also through Paris that Israel financed much of Mossad's work. It was a tricky operation, for the sums involved were considerable, but there were always businessmen to help: one, for example, was a French Jew who made a lot of money speculating in the Paris property boom of the early 1960's. He laundered some of Mossad's money and paid it into the bank accounts Mossad's Paris agents discreetly maintained.

For his part, Aerbel was instructed to take up his life as a European businessman, travelling to and making contacts in as many countries as possible. Mossad also asked him to resume his visits to United States military bases.

Aerbel set about rebuilding his European business. His method was simple: he would contact a company, often out of the blue, and offer his services as a salesman. Job Fass is a Dutchman who owned a factory near Copenhagen that made what are known in Scandinavia as 'Buden' – the term covers souvenirs and knick-knacks of every kind, from trays, dishes and clothespegs to shopping reminders, fondue forks, and model Vikings. 'I was just starting in business,' says Fass, 'when he approached me and asked if he could sell my things abroad. I agreed and I gave him a collection of samples and he took them with him in a suitcase. He travelled around and sent me the orders. I sent him the goods and I got my money. I was very satisfied with him.'

Fass met Aerbel only once or twice a year when he went to Denmark to collect his commission (Aerbel was on ten per cent). Fass was always impressed: 'He was very correct and very precise and he always did the things he promised.' Fass was especially pleased when Aerbel managed to sell some of his goods in the renowned Paris department store, Galeries Lafayette. When Aerbel asked Fass to suggest other companies he could approach, Fass looked some names up in a trade directory and passed them on.

Aerbel's repertoire grew to include furniture, sports equipment, chocolate, and cheese. And he contrived to return to some of the territory he had covered in his days with Danskform. Michael DeWitt is another Dutchman who made wooden Vikings in Denmark. Aerbel approached him in much the same way as Fass. 'He was a very good seller for us,' says DeWitt. 'He sold all over Germany, especially at American bases.'

Although Aerbel consistently won tributes to his selling ability, his personality struck people in different ways. Paul Schlie, a director of the Danish sports company Select, found Aerbel to be a 'floater,' someone who was hard to pin down. 'When we sent him a letter we couldn't really be sure he would answer or say when he would visit us again,' Schlie says. Even so, Aerbel made one proposition that Schlie could not resist. 'He offered to sell our goods in Libya. He said he went there twice a year.' Aerbel did indeed bring Select two orders for footballs, worth some £1,600. But the relationship ended in acrimony when Select, not having heard from Aerbel for some time, started dealing directly with the Libyan buyer. Aerbel resurfaced and demanded an introduction fee, which Select finally paid.

Select was not the only company Aerbel sold for in Libya. When Aerbel made his customary approach to Rudolf Madsen, another Danish manufacturer of souvenirs and household goods, Madsen told him he already had salesmen in Germany and Scandinavia but he was welcome to try his luck elsewhere. Madsen was somewhat taken aback when Aerbel forwarded an order from Tripoli for goods worth £500. 'That was a handsome order for us,' Madsen says. Aerbel came to his office once or twice a year, and Madsen saw yet another side of Aerbel's personality: 'He was a very interesting and funny man. We enjoyed his visits very much. Every time he came he had some amusing thing to tell us.'

Even though Schlie and Madsen were impressed with Aerbel's success in Libya, they were not in a position to appreciate the full extent of his achievement. Aerbel was a Jew, who lived in Israel for eighteen months, and into the bargain was working for Mossad. To have penetrated that far into enemy territory was a remarkable accomplishment that must have brought Mossad considerable fruits. Aerbel later claimed to have made another daring journey in this period, joining Sylvia Raphael, Mossad's 'Canadian freelance photographer', on one of her assignments. He says they cruised on a yacht in the eastern Mediterranean, calling at a number of Arab ports where Aerbel made a note of the vessels at anchor.

But Aerbel was not totally successful at blending the two halves of his double life. His employers found that it was harder and harder to contact Aerbel directly, and they could only leave messages for him in the hope that he would call back. To some he gave his Paris address; to others, the address of an apartment in Rome. He created occasional confusion, too, by using different names – most often Ert but sometimes Aerbel, and occasionally he even reverted to his original family name, Erteschick.

Eventually he seems to have become aware that he was overdoing the air of mystery. In April 1968 he announced to his business contacts a new and permanent telephone number where he could be contacted: Copenhagen 13 10 20. He explained that this was the telephone number of a company he had just acquired. The company's name was Viking; its address, 43A Norregade, Copenhagen. As usual with Aerbel, however, things were not quite what they seemed.

Norregade is a long street that runs north from the city centre. The entrance to 43A is a narrow doorway at the end of a rather gloomy cobbled courtyard. Up a flight of wooden stairs, at the end of a dark and dusty corridor, is a room containing a small upright switchboard and three shelves lined with telephone receivers. Each receiver is marked with a different company name. One man sits in this room answering the calls that come in to those companies.

This is the office of 'Telefon Sekretaeren,' a Danish organization that provides companies with both a respectable address and somebody to take their calls. Aerbel rented one of its telephones in the name of Viking. There was no such company registered in Denmark but that did not deter Aerbel from adding a company nameplate to the collection at the entrance to the courtyard. And whereas most of these were discreetly engraved in metal, Viking proclaimed its existence in large letters painted on a green sign.

Despite Viking's lack of substance, the companies Aerbel dealt with were pleased with the new arrangement. And Telefon Sekretaeren liked their new client. They were used to fending off complainants and creditors but Aerbel gave them no such trouble. He also paid their bills promptly.

By the autumn of 1969 things were going Aerbel's way, and his cover as an ambitious and successful businessman was pretty solid. That is why he must have seemed to Mossad the ideal candidate to take over the affairs of the Biscayne Traders Shipping Corporation. It is, therefore, extraordinary that Aerbel should have jeopardized all that by recruiting someone like Torben Hviid to help dispose of the *Scheersberg A.*

In fairness, however, it should be said that – his criminal propensities aside – Hviid was a perfect accomplice because he displayed a remarkable lack of curiosity about Aerbel's strange life style. Once, for example, Hviid had been astonished when Aerbel turned up in Copenhagen with his hair dyed red, but did not press for an explanation. On another occasion Hviid visited an apartment Aerbel maintained in Rome which was bare and unlived in, and was particularly struck by the lack of any cutlery. Hviid pretended not to notice. (According to the Italian police Aerbel's Rome apartment was used as a Mossad 'safe house' for agents on active service.) And when Hviid agreed to become company secretary of Biscayne Traders he did not ask any questions.

It is also true that Hviid was genuinely fond of Aerbel. Others would have regarded Aerbel as cheap. When, for instance, it was his turn to pay for dinner he would take his friends to an economical vegetarian restaurant where he would extol the virtues of a meat-free diet. When someone else was paying, Aerbel was quite ready to eat an expensive non-vegetarian meal. It was also Aerbel's habit, when calling his wife in Paris from a pay phone, to insert a coin of the lowest possible denomination to allow just time to say, 'Hi Debra, all okay.' But Hviid found Aerbel and all his quirks endearing. Perhaps he recognized in Aerbel some of his own traits: devious ingenuity and an ability to keep any number of balls in play at the same time.

Certainly Hviid was very sorry when in 1971, after the affairs of Biscayne Traders had been tidied up, his friend announced that he and his wife were leaving Paris, and going home to Israel. Aerbel said he felt the time had come to settle down.

Back in Israel Aerbel did two things he had never done before. He took a steady job, and he bought a house. He joined the export division of the Sunfrost frozen foods company of Ashdod, commuting the twenty miles from Tel Aviv each day. And he asked the fashionable Israeli architect Haim Heifetz to build him a home.

Aerbel's years in Europe had not been unprofitable, and he asked Heifetz for something special. Heifetz responded enthusiastically. His design was for a house constructed of three linking domes made of waterproof papier-mache that looked like part of a moon colony.

It had one most unusual feature – almost no interior doors. It was to be built in Herzliyya, the Tel Aviv suburb regarded by many prominent Israelis as the only socially acceptable place to live. (Aerbel was typically eager to advertise his spectacular acquisition. When next in Copenhagen on business he walked into the office of the Danish good living magazine *Bo Bedre* with a set of photographs of the house. *Bo Bedre* took up his suggestion that it should publish an article and an amply illustrated report eventually appeared early in 1973.)

Aerbel had a growing family, two boys and a girl, and he spent most of his time in Israel. But during occasional business trips to Europe for Sunfrost he continued to carry out assignments on Mossad's behalf. In April 1973 Aerbel left Sunfrost and became export director of Osem,* Israel's largest private food corporation. About a month later he heard from Mossad again. There was a job coming up and Mossad asked Aerbel to make himself available. It was fortunate for Mossad that Osem were understanding employers. Even though Aerbel had only just joined the company he was granted a six-week vacation starting on 15 June.

Mossad telephoned Aerbel again on 8 July. The caller was Aerbel's customary control, a Mossad executive named Abraham Gehmer, and he asked Aerbel to come into the

*Hebrew for abundance.

Ministry of Defence offices in north Tel Aviv. There Gehmer told Aerbel that the job was on for the following day. It was connected with 'a threat to Israeli property in Scandinavia, and Aerbel would be away for at most five days.

Gehmer was understating it. The threat was a man called Ali Hassan Salameh and Israel did not have a more implacable enemy.

Chapter 15
The Wrath of Zwi Zamir

Mossad's pursuit of Ali Hassan Salameh reflected a return to one of Israel's earliest preoccupations: revenge. Israel considered Salameh to be a ruthless terrorist. In particular, it held him responsible for the bloody carnage of the Munich Olympic Games in September 1972, when members of a Palestinian movement known as Black September took a group of Israeli athletes hostage. Eleven of the athletes died.

Mossad believed that Salameh was the planning chief of Black September. The movement had been formed in the aftermath of the events of September 1970 when Jordan's King Hussein unleashed his Bedouin troops on the Palestinian guerrillas operating against Israel from refugee camps within Jordan. In the bitter fighting that followed, the guerrillas were all but crushed. At first the new movement concentrated on avenging this act of Arab treachery. On 28 November 1971, Jordan's Prime Minister Wasfi Tal, on an official visit to Egypt, was gunned down in the doorway of the Sheraton Hotel in Cairo. Three weeks later in London Black September tried to murder Jordan's ambassador to Britain. In February 1972 it executed five suspected Jordanian secret service agents in the basement of a house near Bonn.

After this bloody apprenticeship, Black September spread its terror more widely. It blew up oil depots in Hamburg and Rotterdam and bombed a West German factory. The organization also hijacked a Lufthansa airliner whose passengers included Robert Kennedy's son Robert. The ransom extorted for the passengers' release was five million dollars.

Three months later Black September made its first direct attack on Israel. On 8 May four of its members arrived at Tel Aviv's Lod Airport on a flight from Vienna and took their fellow passengers hostage. Israeli commandos disguised as mechanics stormed the plane and killed two of the terrorists, but twenty-two days later at the same airport Black September sympathizers exacted a terrible revenge. Three members of the Japanese Red Army terrorist group, who had been trained in Palestinian camps, attacked the crowded arrivals lounge at Lod with submachine guns and hand grenades. They killed twenty-six people and injured eighty.

In August 1972 Black September succeeded in smuggling a bomb, hidden in a tape recorder, on to an El Al Boeing 707 flying from Rome to Tel Aviv. Although the bomb exploded in mid-flight the intended disaster was averted thanks largely to the skill of the El Al pilot. By that time, however, Salameh was planning the attack in Munich that would take Black September on to the front page of every newspaper in the world.

At 4 am on 5 September eight members of Black September met outside the fence surrounding the village that had been built to house the competitors in the Munich Olympics. Five of the men had travelled to Munich via a circuitous route from Tripoli, Libya. The other three had been in the city for some time: one was an architect who had helped design the village; two had taken menial jobs there to keep an eye on their intended victims. They were all armed with Russian-made Kalashnikov rapid-fire rifles. They were dressed in sweatsuits, so that as they climbed over the fence they were mistaken for recalcitrant athletes coming home from a night on the town. The three who knew the area guided the others to their destination: Connollystrasse 31, where the Israeli athletes were living.

As the terrorists stormed the apartment block they shot dead two of the Israeli team – the wrestling coach and Israel's weight-lifting champion – and took nine others hostage. Those who managed to flee raised the alarm and 500 police converged on the Olympic village. So did the world's press. For the next twenty hours much of the drama was played out on live television.

The terrorists' major demand turned out to be straightforward and impossible. To save the lives of her nine athletes, Israel had to release 200 selected Arab prisoners and deliver them to Cairo. The hostages would also be taken there for an exchange. Israel's policy towards terrorism was, and is, uncompromising. It would not release one prisoner, give one inch. In those circumstances the West German authorities chose to stage a rescue attempt. The plan they drew up was deeply flawed.

In order to lure the terrorists into the open, the Germans agreed to provide them and their hostages with transport to a nearby airfield where a Lufthansa Boeing 727 jet was waiting, supposedly to take them to Cairo. Teams of German sharpshooters were placed along the route and at the airfield in the hope of killing the terrorists before they could harm the hostages. In the early evening of 5 September, Zwi Zamir, head of Mossad, arrived in Munich to offer strategic guidance. He expressed grave doubts about the feasibility of the plan, but there was nothing he could do to stop it.

At just after ten o'clock that night the terrorists and their hostages set out for Furstenfeldbruck airfield, fourteen miles west of the village. They travelled first by bus, then in helicopters, and none of the riflemen hidden along the route had a chance to shoot. But when the helicopters landed at the airfield, and the terrorists stepped out onto the tarmac, sharpshooters opened fire. Their first shots missed. For the next six minutes a firefight raged. When it was over all nine Israeli athletes and five of the terrorists were dead.*

Zamir watched the battle from the airfield control tower. The next day he attended an improvised memorial service for the dead athletes, held in the Olympic stadium. In the section of the stand occupied by the Israeli team, eleven seats were left conspicuously vacant. He took the first flight home to Israel. He was depressed and furious at the West Germans, not only for bungling the rescue attempt but for failing to provide adequate security at the Olympic village in the first place. This was not the first time a European government had made it easy for groups like Black September to operate. Time after time since 1968 there had been Palestinian terrorist attacks in Europe. Yet, in Israel's view, West Germany,

*The remaining three were captured. They were freed six weeks later when West Germany gave into the demands of terrorists who hijacked a Lufthansa Boeing 727 shortly after it left Beirut.

France, Italy and Britain, among others, had refused to take the problem seriously. As a result, eleven Israelis were dead. By the time Zamir's plane reached Tel Aviv he was determined that Mossad would strike out at terrorists wherever they could be found. And it would avenge the dead of Munich.

The concept of revenge was embodied in one of the first laws passed by Israel, the Nazis and Nazi Collaborators (Punishment) Act of 1950. It permitted Israel to try war criminals for acts committed against the Jewish race even before Israel was created. That law's most renowned victim was Adolf Eichmann.

Eichmann was the SS officer who administered Hitler's Final Solution – the extermination of some six million European Jews. In 1945, like many Nazis, Eichmann fled to South America. Twelve years later Mossad received information that Eichmann was in Argentina. Later it learned that he had taken the alias Otto Klement and was living in a scruffy house in a suburb of Buenos Aires. It was not until 1960 that Mossad tracked Eichmann down, and Mossad's boss Isser Harel flew to Buenos Aires to make certain of the final identification. Then Harel did not hesitate, ordering Mossad agents to bring Eichmann back to Israel for trial 'come hell or high water.'

They did. They kidnapped Eichmann in the street, drugged him, smuggled him into Buenos Aires airport and put him on a specially diverted El Al flight to Israel. There Eichmann stood trial for mass murder. He was hanged on 1 June 1962.

This episode marked an important turning point in Mossad's evolution. Although most of Israel's allies sympathized with the motives that lay behind Eichmann's kidnapping they were at the same time deeply alarmed at Mossad's methods. The point was not lost on Prime Minister Ben-Gurion who decreed that Mossad should henceforth employ a little more restraint. And for a couple of years the Institute did behave in a more orthodox fashion – until a man named Wolfgang Lotz pulled off a spectacular espionage coup in Cairo.

Lotz had been recruited by Mossad in 1956, largely because

of his looks. Half Jew, half Gentile, he had been taken to Palestine from Germany by his mother in 1933 at the age of twelve. His first name was changed to Zeev but in appearance he remained notably Aryan. He was spotted by Mossad while serving in Military Intelligence and he readily agreed to take on an extraordinary and dangerous mission. Mossad had long been itching to infiltrate an agent into Cairo and, with Lotz as his candidate, Harel asked the West German secret service, the BND*, for help. The head of the BND, General Reinhard Gehlen, arranged for Lotz to be provided with training and cover. In due course Wolfgang Lotz arrived in Cairo posing as a fervent anti-semite, a horserider and stud farmer and a wealthy man-about-town. The BND's own agent in Cairo was told nothing.

Much helped by his exotic cover, Lotz developed a circle of highly-placed Egyptian friends who were most indiscreet about Egypt's military capability. He also got to know a group of Germans living in Cairo who, Lotz established, were scientists – many of them Nazi exiles – intent on helping Egypt develop ballistic missiles.

When this intelligence reached Tel Aviv it provoked Mossad into taking direct and drastic action. Disregarding the policy of restraint, Harel authorized attacks on the German scientists, their families and their associates. The reprisals began in Switzerland in July 1963 when a light aircraft chartered by a Swiss-Egyptian businessman was sabotaged. The businessman escaped after fortuitously cancelling his flight at the last moment, but his wife and the pilot died in the subsequent crash. In September, a West German who Mossad suspected of supplying Egypt with rocket motors was kidnapped in Munich. Meanwhile, in Egypt itself, the German scientists began receiving letter bombs. One of them exploded and killed five Egyptians.

This campaign of intimidation continued in 1963. In February one of the German scientists, Dr Hans Kleinwachter, left Egypt for a brief visit to Germany. On a lonely road near the Swiss border, Kleinwachter was ambushed by Mossad agents and shot at through the windshield of his car. For-

*Bundesnachrichtdienst.

tunately for him, the bullet was deflected and buried itself in his thick winter scarf.

Mossad's campaign certainly deterred some of the scientists from continuing their work for the Egyptians, but in March 1963 things went badly wrong. On 2 March, Swiss police eavesdropped on a meeting in Basel between two Israeli agents and the daughter of one of the scientists. The agents were arrested and charged with issuing threats.

The publicity surrounding the case was deeply embarrassing to Israel, and Prime Minister Ben-Gurion was furious. Isser Harel argued that the German scientist had, like Eichmann, been a special case and that Mossad's terror tactics were justified, even though they had been exposed. Ben-Gurion disagreed and after a vigorous debate Harel resigned.

That did not altogether end the matter for even after Harel's departure there were still those in Mossad who continued to believe in direct action. In Cairo the gentleman horse-breeder-cum-secret agent Wolfgang Lotz decided to take matters into his own hands and mailed more letter bombs to the German scientists. They failed to go off, were traced back to him, and in February 1965 he was arrested. Once in custody Lotz unsportingly incriminated the BND's Cairo agent Gerhard Bauch even though Bauch knew nothing whatsoever of the affair. Bauch, who had been earnestly building his own relationship with the Egyptian secret service, was promptly arrested and only frantic diplomatic action saved him from standing trial alongside Lotz. He was deported to Germany where he resigned in protest at the BND's perfidy. (Lotz was sentenced to life imprisonment but was released in July 1967 in exchange for Arab prisoners-of-war.)

The entire episode was diplomatically disastrous to all concerned. The West German government was already in deep trouble with the Arab world through the revelation that it had been supplying Israel with arms, and it was furious with Mossad's Major Lotz for making matters worse. It was also furious that Mossad agents had invaded German territory during the intimidation campaign. The Israeli government was at that time negotiating with West Germany over the terms of compensation for lost arms supplies and the affair seriously weakened Israel's hand. Ben-Gurion issued a stern warning that Mossad had no further business engaging in such foolish and open heroics.

Meir Amit, Harel's replacement as head of Mossad, drew the same moral. Mossad had spent years carefully building a cover for Lotz, who had wrecked that work and seriously jeopardized Israel's relationship with one of her most important allies. Lotz had no business out in the open: an agent must live in the shadows. The whole point of a secret service was that it should remain secret. Amit insisted that Mossad should go under cover and stay there, and his eventual replacement Zwi Zamir pursued the same policy.

Plumbat was the perfect example of that concept: a brilliant operation carried out in secret which had remained a secret. Then came Cherbourg – Mossad's design had held until the last minute, when the complication of the part played by the Israeli Navy forced Israel out into the open.

But after the Munich massacre Zamir was determined that Mossad must, once again, change the rules.

Even before Munich, Zamir had proposed that in the fight against terrorism Mossad should take the law into its own hands. In the summer of 1972 he had told the inner cabinet that Mossad should set up assassination squads to carry the war wherever Israel's enemies could be found. Zamir argued that it was a logical extension of Israel's existing policy of mounting reprisal raids against Palestinian leaders in Lebanon. In retaliation for the massacre at Lod airport, for example, Israeli frogmen went ashore in Beirut and planted a bomb in the car owned by Ghassan Kanafani, spokesman for the militant Popular Front for the Liberation of Palestine (PFLP). Both Kanafani and his teenage niece were blown to pieces. But before Munich the cabinet had baulked at the idea of direct action in Europe. Killing people on Arab territory was one thing; murdering them in London or Paris, quite another. In Munich's emotional aftermath such prosaic diplomatic considerations carried less weight. On 6 September, only hours after returning to Israel from Munich, Zamir argued his case again. This time the inner cabinet agreed. Salameh and his ilk should be hunted down and killed wherever they were. Prime Minister Golda Meir gave Zamir the go-ahead to set up the killer squads.

The Palestinians were not to die quietly. Israel wanted them killed in spectacular fashion as a clear warning to others. For Mossad that posed the problem of who should actually pull the trigger or plant the bomb. It had discovered during the Wolfgang Lotz affair that secret agents who engaged in assassination attempts tended to wreck their carefully constructed cover. As Mossad was unwilling to jeopardize its European network of agents the murders would have to be carried out by others.

The solution was to create a self-contained operations squad, under Mossad's control. The men and women in this squad would be the core of the 'hit teams' – the killers.

The job of Mossad's resident agents was to locate the targets, provide information and logistic help and then largely stay out of the way when the 'hit team' came in, killed and got out fast. It was a fine distinction. Some of Mossad's resident agents in Europe became heavily involved in the murders and the network was put at risk.

Within a month of the inner cabinet's decision, the revenge teams* were ready to go. Ali Hassan Salameh was the most wanted man on Mossad's list but he was proving elusive. He was believed to have followed the events in Munich from East Germany and then crossed to the west where he dropped out of sight. As the hunt for Salameh went on Mossad turned to easier targets.

The first to die was Wadal Adel Zwaiter, suspected by Mossad of being the leader of Black September's Italian cell. At about 10:30 pm on 16 October 1972, he returned to his apartment on Piazza Annibalino in Rome. As he entered the building he was shot twelve times at point blank range by two Mossad killers armed with handguns.

The same month, in Paris, Mossad killed a Syrian journalist named Khodr Kannou who was suspected of being a member of the PFLP. He, too, was shot down outside his apartment.

Mahmoud Hamshari was next to go. He was the Paris spokesman of the Palestine Liberation Organization but

*Some observers dubbed them Wrath of God (WOG) squads.

Mossad regarded him as a terrorist. In particular he was reckoned to have inspired the planting of a bomb on a Swissair flight to Tel Aviv which had cost the lives of forty-seven people. On 7 December he was lured away from his apartment at 175 Rue Alesia while a bomb was planted in his bedroom. The next day he answered his telephone and heard a high-pitched noise. It was an electronic signal and it triggered an explosion that shattered the lower half of his body. He died a month later.

Six weeks after the attack on Hamshari, Mossad struck in Nicosia, Cyprus. The victim this time was Bashir Abu Khair who Mossad believed had gone to collect a regular donation to the Palestinian cause from the Russian Embassy in Nicosia. Khair returned to his room at the Olympic hotel, got undressed and went to bed. A bomb, hidden under the mattress, was triggered electronically and blew him to pieces.

In April two more alleged terrorists died. Professor Basil Al Kubaissi of the American University in Beirut was believed by Mossad to be in charge of terrorist arsenals in Europe. A killer squad caught up with him in Paris on 6 April: as he walked to his hotel, two young men overtook him and shot him nine times. On 9 April, Zaiai Muchasi, believed to be a replacement courier for Bashir Abu Khair, was dispatched in his Nicosia hotel, again with the aid of a bedroom bomb.

In June Mossad claimed its most important victim so far. Mohammed Boudia was an Algerian who lived in Paris where he had gained a considerable reputation as an actor and theatrical director. He was also a dedicated terrorist and head of the European cell of the PFLP. He had masterminded several attacks against Israel and had liaised with the Japanese Red Army before its bloody assault on Lod Airport. On 28 June Mossad took its revenge. After spending the night with a girlfriend, Boudia drove his car to the Rue des Fosses-Saint-Bernard and parked. While he was away, a Mossad assassin planted a powerful explosive charge behind the driver's seat. It was fitted with a short fuse that was triggered when Boudia opened the car door. He just had time to get into the driver's seat before the bomb exploded.

This murderous campaign had not gone unanswered. In Washington DC, Colonel Yosef Alon, air attache at the Israeli

Embassy, was shot dead in the garage of his house. Israeli diplomats around the world received a flurry of letter bombs. And Israelis in Rome and Nicosia who were believed – probably wrongly – to be Mossad agents were gunned down. Mossad did lose a man in Madrid, early in 1973. Moise Yshai was an agent who thought he had successfully infiltrated Black September's Madrid cell but his real identity had been discovered. At about 10:30 am on 26 January, Yshai was struck by five bullets as he walked along one of Madrid's main streets, intending to keep an appointment at the Morrison cafeteria. His killer, a medical student in his mid-twenties from Jordan who had arrived in Madrid a week earlier, was never caught.

But Mossad was winning. Leading terrorists went into hiding and the number of their attacks in Europe dropped dramatically. In nine months the 'hit teams' had carried out seven assassinations without getting caught. The police, particularly in Paris, had little doubt who was behind the killings, but it was impossible to prove any link with Israel. Mossad's network in Europe remained intact.

And in June 1973 Mossad picked up the trail of the man who had always been the prime target: Ali Hassan Salameh. He was spotted in northern Europe, heading for Scandinavia. In Israel a team was assembled from available agents to take up the chase. One of those agents was Dan Aerbel.

Chapter 16
A Fatal Mistake

On 9 July 1973, Aerbel dutifully followed the instructions of his Mossad control and presented himself at Lod Airport, near Tel Aviv. There he met two men who introduced themselves as Gustav Pistauer and Jean-Luc Sevenier. Pistauer was one of Mossad's most experienced agents. Originally from Poland, he had spent much of World War Two in German concentration camps, where many of his relatives had perished. As well as Hebrew and Polish, he spoke English perfectly, and was accomplished in German, French and Italian. He was in his forties, and close to Mossad's retirement age. His companion Sevenier was a newer Mossad recruit. Some twenty years younger than Pistauer, he had come to Israel from one of France's former colonies in North Africa.

That afternoon the three men caught a Lufthansa flight to Frankfurt, en route to Denmark and then Sweden. Mossad had received reports that Salameh was moving north through Europe towards Scandinavia, and felt that his most likely destination was Stockholm, which at that time had the unenvied reputation of being the European base of several terrorist groups. The three men's task was to try to find Salameh. If they succeeded, an assassination squad would be brought in to kill him. But Pistauer told Aerbel none of this: only that they were bound for Scandinavia, where Aerbel's languages would be required. When Aerbel started probing further Pistauer told him bluntly to stop asking questions.

After a night in Frankfurt the three men flew to Denmark, where they were due to change planes. From the transit lounge Aerbel telephoned his sister, a doctor with a practice in Copenhagen, to say hello. He told her he was on his way to an international conference on vegetarianism in the province of Skomo in southern Sweden. The purpose was wrong; the country right: later that day the trio arrived in the Swedish capital, Stockholm.

They were there for a week, and if his two taciturn colleagues were uncovering clues to the whereabouts of Ali Hassan Salameh, Aerbel knew nothing about it. Pistauer instructed him to rent two apartments for six months. Aerbel had no idea why they were needed for so long: 'I had to equip the apartments with furniture,' he said later. 'I also had to get hold of a lot of keys.' Pistauer made Aerbel read aloud the political news from the papers each morning, and generally treated Aerbel as his amanuensis. Aerbel was surprised that Pistauer seemed so preoccupied with his forthcoming pension, calculating aloud how much money he would receive. He found Pistauer hopelessly impractical. 'I had to help him with everything,' Aerbel said.

On 13 July Aerbel called his wife in Tel Aviv to tell her he would be home soon. On 16 July he was still in Stockholm. 'I said in a very decisive way that I wanted to go back to Israel. "No," Pistauer answered. "We are going to Norway."'

The reason for the switch in plan was new and better information that had reached Mossad from Switzerland. An Arab named Kemal Benamene had been seen by an Israeli agent at Geneva airport buying a ticket for Oslo. Mossad believed that Benamene was an important courier for Black September: if he was going to Norway, then there was a strong chance that Ali Hassan Salameh was already there. The three man advance team of Pistauer, Sevenier and Aerbel was dispatched to Norway to locate Benamene and keep an eye on him. Meanwhile, back in Israel, Mossad assembled an assassination squad.

For Mossad both the abruptness of Salameh's reappearance and his apparent location were awkward. Previous assassinations had been carried out in countries where the preparations had been thoroughly laid. In Paris, scene of more than half the murders, Mossad had been operating on what was virtually home ground, and the 'hit teams' had been able to rely on the well-established resident agents for logistic support. In Norway, Mossad's network was sketchy, and there the 'hit team' would have to fend for itself. In view of that it would have made tactical sense for Mossad to do no more than track Salameh to more suitable territory, or, at least, delay the operation until the groundwork in Norway had been properly laid. It did not because killing Salameh had become an obsession. Just as though it were back in the old days of brilliant improvisation, Mossad hurriedly formed a 'hit team' from the best recruits it could find and packed them off to Norway. One of those recruits was very raw indeed.

Marianne Gladnikoff was a short and slightly plump Swedish girl of twenty-nine, with brown eyes and long black hair. She was a doctor's daughter from Stockholm who had emigrated to Israel out of feelings of solidarity after the 1967 Six Day War. She had a degree in languages from Stockholm University but found no suitable jobs in Israel and took a course in computer programming. She shared an apartment with a girl friend in the suburb of Bat Yam south of Tel Aviv, and worked as a programmer for Electronics Industries of Ashdod, known as ELTA. ELTA undertook sensitive work for Israeli government agencies, including the preparation of programmes for Mossad's computers. Gladnikoff was warned not to tell anyone that she worked on computers and to describe herself as a secretary. Early in 1973 she had applied for Israeli citizenship, and on 2 July she started a top-secret induction course that would eventually qualify her to join Mossad.

Just two weeks later, Aerbel's control Abraham Gehmer called Gladnikoff and asked if she was prepared to go on a mission for Israel. She said she was and Gehmer told her to go to a Tel Aviv cafe named the Angel. There she was met by a man she did not know. He told her that 'nothing illegal' would be required of her, and took her passport, saying it was needed to make out her air ticket – where to, she had no idea. In less than ten minutes the meeting was over.

Early the next morning, 18 July, a mini-bus picked up Gladnikoff at her apartment and took her to Lod Airport. Some fourteen or fifteen other people were gathered there, including three beautiful women. Gehmer was the only one she knew. They spoke in Hebrew and as she understood only a smattering of the language she could glean little of what was going on. But she was particularly struck by one self-possessed man in his forties who everyone addressed as Mike. Had she known his true identity she might have realized that she was involved in a critical assignment. Although he had adopted the alias of Edouard Lasquier for this job, his real name was Georg Manner. He was the head of Mossad's operations in Europe.

Manner handed Gladnikoff a large roll of banknotes, including a thousand US dollars – Mossad never stints on expenses – and directed her to a plane. The team had been divided into two groups, both bound for Oslo, one via Zurich, the other via Amsterdam. Gladnikoff was in the Zurich party, and sat next to one of the beautiful women, who happened to be Sylvia Raphael. Raphael said her name was Patricia Roxburgh but said little else during the whole flight. In Zurich, one of the men, who said his name was Jonathan Ingleby, discovered he had lost his baggage. Gladnikoff rode with him into Zurich between flights, and they both spent some of their Mossad dollars on new clothes. The last person in the Zurich group was Gehmer, who had asked both Gladnikoff and Aerbel to join the mission. As the plane touched down in Oslo, Gehmer told Gladnikoff that she was to be used as a translator and interpreter. She could not speak Norwegian but she assumed that, with her Swedish, she was the best that Mossad could find. In Oslo she was pleased to discover someone else she knew. It was Aerbel, who had once come to the ELTA office to seek her help in translating some Scandinavian documents.

With the advance guard of Aerbel, Pistauer and Sevenier, who had been in Oslo for twenty-four hours, a sizeable team had now assembled. A number of them had previously worked for Mossad in Paris, including all three who had flown in with Gladnikoff via Zurich. Sylvia Raphael, of course, had been based there as the photographer 'Patricia Roxburgh.' Jonathan Ingleby, formerly Jonathan Isaac Englesberg, the British businessman specializing in East-West trade, was a trained marksman who had been in Paris at the time of several earlier 'hits,' and now in Norway wore his .22 Beretta in a shoulder holster. Gehmer had been attached to the Israeli Embassy in Paris from 1965 to 1969, appearing in the official diplomatic list. In Norway Gehmer was posing as the Englishman Leslie Orbaum, and carried a British passport in that name, number 604285. It was an example of Mossad's technique of providing agents with the identity of real people who were unaware their names were being used. In this case the real Leslie Orbaum was an Englishman who worked as a school teacher in Leeds. The real Gehmer was born in Rehovot, in Palestine, in 1937.

There were several agents with similar cover in the group that had flown in via Amsterdam. There was Gerard Emile Georges Lafond, for example: the real Lafond was born in Bourg, France, on 13 February 1945, lived for a time at 144 Rue Lafayette, Paris, and emigrated to Israel in July 1971, going to live and work on the Kibbutz Ein Hachoresh near Haifa. The Mossad agent now in Oslo had borrowed his name without asking. A second in that category was the agent calling himself Rolf Baehr, who was a Mossad wheelman, or driver. The real Rolf Baehr was born in Cologne on 11 February 1930, and had fled with his parents to Palestine via Italy to escape Nazi persecution. Another agent with a Paris background was Zwi Steinberg, who, like Gehmer, had been officially attached to the Israeli Embassy, from 1968 to 1971, before setting up in Paris as a gentleman of leisure, and as such received a nice round 3,000 francs (£250) a month in his account at the Credit Commerciale de Paris. Steinberg had the advantage of dual nationality, and in Norway he used his Brazilian passport.

Another agent with a South American background was Raoul Cousin, from Argentina. And completing the team were two women: Nora Heffner, a tall, attractive thirty-year-old; and the strikingly beautiful twenty-five-year-old Tamara, who also sometimes called herself Tamar, sometimes Marie. She was one of Mossad's professional killers and the mistress of Georg Manner, Mossad's European boss. Manner was in Norway to lead the operation and Pistauer was his deputy.

Aerbel was still finding the work Pistauer was assigning him less than enthralling. He had made all the travel arrangements to get the advance team to Oslo – 'Pistauer couldn't even arrange his own air ticket,' he said later – and now in the name of Dan Ert he rented two cars and an apartment in Otto Ruges vei* in the residential suburb of Baerum. But he did begin to divine something of the nature of the mission when Pistauer told him to call all Oslo's hotels in search of an Arab named Benamene. Aerbel felt very pleased when he found him at the Panorama Summer Hotel, one of the student hostels in Oslo which doubles as a hotel during the long summer vacation. But the 'hit team' delayed going there for several hours, by which time Benamene had moved on. That evening, while Pistauer and one of the new arrivals from Tel Aviv enjoyed a meal at the Oslo Tivoli grill, Aerbel made some twenty more calls from a nearby pay phone in a vain effort to find Benamene again.

On 19 July the character of the mission, and the parts played by Aerbel and Gladnikoff, changed decisively. It became an uninhibited manhunt. One of the team managed to trace Benamene to the small town of Lillehammer, an inland resort 120 miles north of Oslo. Manner reasoned that Black September's courier could only have gone there to meet Ali Hassan Salameh, and no fewer than ten members of the 'hit team' sped north in four rented cars. Among them were Aerbel and Gladnikoff, who now must have realized that they were involved in something very serious.

The team arrived in Lillehammer to find that Benamene was staying quite openly at the Skotte, a small tourist hotel in the town centre. Gladnikoff and Cousin were assigned to stay there in order to watch him, while the eight other agents checked in at a variety of hotels. For someone on so important a mission Benamene was certainly relaxed. That afternoon he went for a swim, then returned to the Skotte for a sleep. Unsure where he was, Cousin and Gladnikoff took up a position in the Skotte's television lounge so they could see if he went in or out. After three hours Benamene came downstairs, and, to their considerable alarm, sat down beside them. Cousin and Gladnikoff did not dare move until half an hour later when Benamene stood up, stretched and went back to bed. Cousin and Gladnikoff hurried out to report to Manner.

*Vei is Norwegian for way or street; Otto Ruges was a Norwegian general of World War II.

In the morning, things hotted up. Benamene went for a walk. Cousin sent Gladnikoff after him; she lost him. The 'hit team' split up, found Benamene sitting alone at an outdoor cafe, and then lost him once more. He was not spotted again until midday when he sat at the Karoline cafe on Lillehammer's main street, Storgaten. This time he was talking to two men – and one of them was an Arab!

The vital task of identifying Ali Hassan Salameh was entrusted to Pistauer, Cousin and Gladnikoff. They went boldly up to the Karoline street cafe and occupied a table barely five yards from their quarry. Pistauer was equipped with a photograph of Salameh, an enlargement from an outdoor colour snapshot in which Salameh was pictured from the waist up. The enlargement had rendered the image somewhat indistinct, and care was obviously needed. Pistauer looked at the picture, then stared directly at the Arab sharing Benamene's table. Gladnikoff wondered why Benamene remained so composed: didn't he recognize her and Cousin from the night before?

Apparently not. Benamene and his companions went on with their conversation, even though it was fifteen minutes before Pistauer made up his mind that the second Arab was indeed Salameh. The identification took that long because the photograph, besides being indistinct, showed a man who was clean-shaven, whereas the Arab with Benamene wore a moustache. But if Pistauer needed confirmation that he was observing a meeting between Salameh and his courier it came when he saw the two men exchange handwritten messages.

Having found Salameh the 'hit team' let him go. When the two Arabs eventually got up from their table, they shook hands and parted: Benamene walked off in one direction while Salameh rode away on a bicycle in another. As the team did not have bicycles they followed Benamene. They lost him. Fortunately, Manner anticipated that Benamene would return to Oslo and he dispatched Aerbel, Pistauer, Gehmer and Raphael to the capital in the hope of picking up his trail. The four agents arrived in Oslo in time to see Benamene arrive by train and they followed him around the city for several hours. In the process, however, Pistauer managed to pull a leg muscle and retired from the hunt to a bed in the Ritz Hotel.

Late that night Manner sent a message to Oslo: Salameh had been located again. Aerbel, Gehmer and Raphael left Pistauer to his own devices and headed back to Lillehammer. Mossad was closing for the kill.

At ten o'clock the next morning, 21 July, the team gathered in the parking lot of Lillehammer's railway station. Manner was determined not to let Salameh slip away and he ordered all of the agents to join in tracking the target. They used the four rented cars and depended for communication on walkie-talkie radios: to improve reception it was sometimes necessary to operate them with their aerials sticking out the car windows.

The team picked up Salameh's trail at an outdoor cafe just off Storgaten. When he left the cafe Gladnikoff and Cousin were assigned to follow on foot. In the busy Saturday morning throng on Storgaten they temporarily lost sight of him, but Lillehammer is hardly overcrowded with Arabs and it was not long before the rest of the team relocated him. Salameh led them to a swimming pool.

Manner ordered cars into positions overlooking all the pool exits and instructed Gladnikoff to follow Salameh inside. One agent bought her a bikini which proved much too small, so she rented a one-piece suit at the ticket office. In the pool she saw Salameh speaking to a European. Cautiously she swam closer, only to discover that they were talking in French, which she could not understand.

When Salameh left the pool he and the European walked to another cafe on Sorgaten, followed by Cousin, Raphael and Aerbel. Salameh stayed there until 12:30 when he was joined by a woman with blond hair who was obviously pregnant. They got onto a bus, rode to a residential area on the edge of Lillehammer and disappeared into an apartment building. All four of Mossad's rented cars followed the bus and then took up positions around the building.

The 'hit team' waited over six hours for their quarry to reappear. To relieve the monotony, the watching agents sometimes switched cars, or parked close to each other so that they could chat. Occasionally, they took spins around the neighbourhood. Once a police car passed by when two of the cars were parked close together, but appeared to notice nothing untoward. Around seven o'clock most of the watchers drove back into Lillehammer to have dinner, leaving Gladnikoff and Cousin watching the main entrance. That was unfortunate for soon afterwards Salameh and the pregnant woman, now wearing a bright yellow raincoat, walked out of the apartment block and started downhill towards the centre of town. There was a moment of understandable panic. Cousin ordered Gladnikoff to drive into Lillehammer to find Manner, while he followed Salameh on foot. She refused because she did not know how to drive the car. Salameh and the woman walked past as Gladnikoff and Cousin were at the height of their quarrel. Cousin got out, leaving his fellow agent to cope as best she could. Fortunately for her, she came across Manner, now accompanied by Tamara, almost immediately. And fortunately for the whole team, Cousin managed to stay with Salameh and the pregnant woman. He followed them to the cinema, watched them go inside and then reported to the station parking lot which had been selected as the emergency rendezvous.

Aerbel, Gladnikoff and Gehmer were instructed to keep watch on the cinema's entrance. Raphael and Cousin sat in two cars close by. Meanwhile, the assassination squad prepared to go into action. The two killers were to be Jonathan Ingleby and the young woman Tamara. Their driver was Rolf Baehr and the fourth in the squad was Gerard Lafond. They drove in a white Mazda to a point a quarter mile or so from the apartment building where Salameh had spent the afternoon and waited.

When the movie, *Where Eagles Dare*, ended shortly after 10:30 Salameh and his companion left the cinema and walked up Storgaten. Aerbel, who spotted them first, went after them, followed by Gladnikoff and Gehmer. Raphael and Cousin went ahead in their cars and waited further up the street. When Salameh boarded a bus Cousin used his walkie-talkie to warn the assassination squad.

Baehr drove the Mazda to a point close to the apartment building and waited for the bus to arrive. When Salameh and the woman climbed down from the bus Baehr eased the car downhill towards them. He passed so close that the car almost touched Salameh. Then he jammed on the brakes. Ingleby and Tamara got out and raised their Berettas. Ali Hassan Salameh only had time to say 'No.'

It was the style of the Israeli 'hit teams' to leave nothing to chance. When Wadal Zwaiter was assassinated in Rome he was shot twelve times at point blank range. In Paris, Basil Kubaissi was shot nine times. In Lillehammer, Salameh took fourteen bullets. The first six tore into his stomach. The next two struck his head, one ricocheting into a nearby house, the other penetrating below his ear and hitting the base of his brain. Shots nine through fourteen ripped into his back as he lay on the ground. Two of these struck his left forearm, one a buttock and others hit his kidney and liver. Even so, Salameh did not die immediately. But when Manner arrived on the scene in a green Volvo a few minutes later he was satisfied that his victim could not possibly survive long enough to reach a hospital.

It was an utterly ruthless killing – exactly the fate Mossad had decreed for Ali Hassan Salameh. The trouble was, the 'hit team' had just killed the wrong man.

Chapter 17
Confession

It is hard to see how anyone could have mistaken Ahmed Bouchiki for Ali Hassan Salameh. The most cursory inquiries would have revealed that Bouchiki, the man Mossad killed, was a well-known Lillehammer figure who worked as a waiter in a health clinic and was married to a local girl named Torill. Aged twenty-nine, with an Algerian father and a Moroccan mother, he had first come to Norway in 1965. He had taken a Norwegian government course in the hotel trade and had worked as a waiter, with a couple of interludes, ever since. He met Torill, an employee at the Lillehammer hospital, in 1972 and they were married later that year, with Torill two months pregnant. To improve his prospects Bouchiki was training as a lifeguard and swam frequently at the local pool. It was there, on 19 July, that he happened to meet Benamene, whom he had not previously known. This chance encounter with a stranger was to cost Bouchiki his life.

There are indeed few Arabs in Norway, and Bouchiki was delighted at the chance of conversing in his own language. The acquaintanceship was renewed at the Kroline cafe the following morning, when they talked about clothes and Arab music. At the end of the conversation, so closely observed by Pistauer, Cousin and Gladnikoff, they exchanged names and addresses. It was this casual action on which Pistauer placed so sinister an interpretation and which convinced him that Bouchiki was their target, Salameh. From that moment Bouchiki was marked for death. He was followed remorselessly the next day: as he went swimming again, and afterwards met Torill, by now seven months pregnant; as they returned to their apartment for the afternoon; as he escorted her, in her bright yellow raincoat, to the cinema that evening; and as they took their last journey home together that night.

Of all Mossad's mistakes in Norway, the botched identification was the most extraordinary. Mossad had spent three years making sure it had the right man before kidnapping Eichmann from Argentina. Pistauer took just fifteen minutes to sentence Bouchiki to death. Marianne Gladnikoff said later that he did so on the basis of the one enlarged and fuzzy colour snapshot which he examined at the Karoline cafe.

It was true that there was a certain facial resemblance between Bouchiki and Salameh – they were after all both Arabs. But Gladnikoff claimed that both she and Cousin had voiced doubts which Pistauer had overridden. She said she had expressed unease about the matter of the moustache – Bouchiki had one at the time; Mossad's photograph showed Salameh clean-shaven – as well as a clear difference in the shape of their eyebrows.

But Mossad should not have relied solely on a photograph. Salameh has several features which could have served as an aid to identification. For example, he bears a distinctive scar or pockmark on his left cheek. And one further characteristic possessed by Salameh is so striking that Mossad's ignorance on the point is astonishing: his height. By all standards apart from American basketball players Salameh is very tall – at least six feet three inches. Bouchiki was five feet seven.

There were other factors that Mossad should have heeded. Would Israel's most wanted enemy, who even in Beirut is constantly attended by a bodyguard, really have behaved so casually? Would he not have been alerted by the people tailing him so conspicuously? The mistake was so outrageous that Israel has encouraged several equally outrageous theories which attempt to minimize it. One was advanced by lawyers defending the arrested Mossad agents at their trial in Oslo in January 1974. They suggested that Bouchiki was a 'sleeper', a Palestinian agent planted many years before and waiting for the call to duty. It is a theory entirely without proof: if Bouchiki, a Moroccan, was involved with the Palestinian movement, not one scrap of evidence to show it has emerged to this day. Nor does the theory get round the problem that in any case Mossad thought it was killing Salameh.

An even more far-fetched theory has emerged: Mossad was set up. According to this theory, Fatah, the military wing of the Palestine Liberation Organization, deliberately led Mossad to the innocent Bouchiki in order to divert attention from the Arabs' preparation for the October 1973 war. For Israel, the theory has the advantage of excusing both the killing in Lillehammer and Israel's lack of preparedness for the war. If it is true, then Fatah pulled off one of the most spectacular subterfuges in the history of intelligence operations. And both Mossad and Israel stand exposed as among the most spectacular dupes.

The truth is much more simple. Mossad's behaviour in Norway was that of an organization that had grown complacent, even arrogant, in its prowess. Mossad was careless of the hazards inherent in operating on unknown territory, instead of the comforting surroundings of a city like Paris where the backup was strong and the ground well-prepared. To arrive team-handed in the small town of Lillehammer was reckless. Local people soon noticed the squad of cars with Oslo license plates, and several jotted down their numbers. The cars' movements on 20 and 21 July were well observed, as were the radio antennas protruding from their windows. The Lillehammer police did see the stakeout of the Bouchikis' apartment on the fatal afternoon. A neighbour was even watching from close by when Bouchiki was gunned down, and noted the type of car the killers had used. By the time the 'hit team' left Lillehammer for Oslo the police of Norway were already searching for them.

A murder anywhere in Norway is a rare event – there are only about thirty a year. The Lillehammer police had dealt with one back country killing in the 1960's, but there had not been a murder within the town limits for forty years. They were on the scene within moments of receiving calls from the Bouchikis' neighbours. Bouchiki may just have been alive when an ambulance picked him up, but at the Lillehammer hospital doctors certified him 'dead on arrival.' News of the killing was quickly flashed to Norwegian police headquarters in Oslo.

Meanwhile, the four Mossad cars, having received their 'go home' instructions from Manner, stopped at the roadside rendezvous five miles south of Lillehammer. According to Gladnikoff, it was Jonathan Ingleby, when asked how the killing had gone, who uttered the laconic 'a job is a job.' Despite Ingleby's studied nonchalance, the atmosphere in the white Peugeot containing Gladnikoff and Aerbel as it continued on the way to Oslo was anything but relaxed. Gehmer, who was driving, said nothing. Beside him Raphael smoked incessantly and gulped whisky from a bottle of Chivas Regal. Aerbel was clearly feeling under pressure, for he swigged whisky, too, despite his claims to be teetotal. He also complained of stomach ache. Next to him Gladnikoff, finally aware, if she was not before, just what the mission was about, was moaning softly and Aerbel took her hand. (Later, he told the police that this was to comfort her. But when he went on trial Aerbel, in a show of bravado, said he was making a pass at her.)

As the Peugeot sped south along the eastern shore of Mjosa Fiord, several cars bearing detectives from Norway's serious crimes squad, the E-Gruppa,* were racing to Lillehammer by the shorter, western route. Local police along both routes had also been alerted, and they set up roadblocks. The first in the Peugeot's path was at Hamar, forty miles south of Lillehammer. It was not a very intimidating affair; a patrolman stood on the roadside signalling cars to stop. The first drivers to approach simply ignored him and drove past. Gehmer was betrayed by his own anxiety. When he saw the policeman he slowed, then changed his mind, accelerated and drove on. His hesitation was enough to alert the patrolman who noted the Peugeot's license number, DA 97943. It was relayed to Oslo police headquarters.

The Peugeot drew up outside the apartment Aerbel had rented in Otto Ruges vei, Oslo, in the early hours of Sunday morning 22 July. The four agents went inside. They were hungry and wanted to eat, but the apartment contained no food so they went to bed.

*Etterforsknersgruppa – the equivalent of a homicide department or the British murder squad.

Later that morning another of the Lillehammer team arrived at the apartment in the green Volvo that Manner had used during the operation. The Volvo, which had been rented from Oslo's Fornebu Airport, was no longer needed, and Aerbel volunteered to take it back, offering to buy some food at the same time. The combination of parsimony and hunger was the 'hit team's' undoing.

As Aerbel drove the Volvo to the airport, Gladnikoff followed in the white Peugeot in order to bring him and the food back. Aerbel parked at the airport terminal and went inside to hand in the keys and to buy food at the terminal's self-service cafeteria. Gladnikoff waited in a no waiting zone. The Peugeot's number, picked up at the Hamar roadblock the previous night, had already been circulated to the police and airline staff at Fornebu. An SAS airline ticket clerk spotted the Peugeot and called a police officer, who took Gladnikoff to the airport police post. As Aerbel walked back towards the Peugeot, his arms laden with food, he was met by more police.

During those first moments in custody, the naivete of Aerbel and Gladnikoff was pitifully exposed. When Aerbel was asked why he was carrying so much food, he explained that some was for his friends. At nearby Sandvika police station, Aerbel said those friends were staying with him in Baerum. Gladnikoff even gave the address of the 'safe house' – 77c Otto Ruges vei. The police went there and arrested Raphael and Gehmer. By 6:30 that evening all four of the Peugeot's passengers were back in Lillehammer, this time in the police station on Storgaten, the same street where the fatal identification of Ahmed Bouchiki had been made.

Some very puzzled members of Norway's E-Gruppa were waiting for them. They were led by Magnhild Aanestad, a detective with slender wrists, greying hair and a taste for simple summer dresses. Aanestad was the E-Gruppa's second-in-command, and the highest-ranking woman on the Norwegian police force. She joined the police in 1954 and even though women were not accorded equal promotion until 1960, rose steadily to *politiforstebetjent*, parallel to a US captain of detectives or a British chief inspector. Despite her frail, almost innocent look, she had tackled and solved her fair share of the E-Gruppa's toughest cases, including a particularly intractable murder involving a Pakistani whose body was found six months after it had been thrown into a fiord. She occupied a large, drab office in the E-Gruppa's headquarters in Oslo's Victoria Terrasse, relieving its institutional greyness by bringing in flowers – she liked Michaelmas daisies – which she displayed in a slender vase on the windowsill. She had no illusions about the requirements of her chosen profession. She remained unmarried, accepting that no husband would easily tolerate the abrupt departures and the long absences. She had been asleep in her Oslo apartment when the news from Lillehammer came through. She was up and dressed in moments, and on the road to Lillehammer within an hour.

The few details Aanestad had learned simply did not add up. The police had heard from a distraught Torill Bouchiki how a man and a woman she had never seen before had burst from a car to shoot her husband down. Aanestad wondered why strangers should come to Lillehammer to kill an Arab. Her first thought was that it must be something to do with narcotics. As she considered the four suspects, two men and two women, who had been brought from Oslo, she felt even more baffled. Two seemed fairly composed. The other two, Marianne Gladnikoff and the man who said his name was Dan Ert, were clearly very nervous.

It fell to Aanestad to assign a detective to each suspect. She had an experienced and eager team available, including a young detective inspector named Steinar Ravlo. Only twenty-nine, Ravlo had already helped investigate some thirty murder cases – quite a tally for Norway. His reputation lay not only in his skill as an investigator, but in his ability as a thoughtful and determined interrogator. In that role Ravlo was undoubtedly aided by his handsome Nordic looks and his deceptively friendly smile. Aanestad decided that Ravlo was the obvious choice for the uneasy Dan Ert. To Marianne Gladnikoff she assigned a short, young, strongly built detective named Leif Lier.

Ravlo escorted Aerbel to an office on the third floor of the Lillehammer station with windows overlooking Storgaten. The room was well-appointed by the standards of Victoria

Terrasse with a splendid working table made of polished mahogany, and a comfortable visitor's chair. Ravlo invited Aerbel to sit down.

When it came to composing a cover story, Aerbel, despite his lack of preparation, had much in his favour. He had travelled widely and he still owned the Viking company in Copenhagen, even if it was little more than a nameplate outside a scruffy office. He also had his fertile imagination.

Aerbel repeated that his name was Ert and said that he lived in Rome. He owned a Danish company named Viking, and he had come to Norway to sell his products. That was why he had rented the apartment in Baerum. He had first met his three companions at an Oslo hotel, and Patricia Roxburgh* had moved into the Baerum apartment with him soon afterwards. On Saturday they had decided to go for a drive, and they could possibly have been in Lillehammer. 'But really I'm not sure at all where we went. We were only looking at the beautiful countryside and we did not have any definite travel plans or schedule.' The four had spent the night in the apartment and in the morning, as the police well knew, he and Gladnikoff had gone to Fornebu Airport to return one of their rented cars. He knew little about his companions' background, but he had noticed nothing strange about them. He definitely knew nothing about a killing in Lillehammer.

The four E-Gruppa interrogators were sequestered with their suspects in different parts of the borrowed Lillehammer police station. Magnhild Aanestad had set up a 'murder room', and from time to time the interrogators would come out to relate to her what they were being told. Despite some inconsistencies, Aerbel's story was broadly similar to those being told by Raphael and Gehmer. Aanestad felt they were all probably lying, but just how they were connected with the murder, or with each other, remained a mystery. When one official arrived at the police station and asked what was going on Aanestad replied: 'A man is dead, that is reality. What the motives are, we don't know.'

But the E-Gruppa detectives were about to get their first

*The alias of Sylvia Raphael.

intimation that they were caught up in an international intrigue with sensational implications.

Gladnikoff had been closeted with Leif Lier for a long time. She was both bewildered and a little angry. She had been kept ignorant, even misled, over the purpose of the mission to Norway. Now she found herself caught up in a murder. Under gentle, almost sympathetic pressure from Lier, she began to talk. She said she had come from Israel, and that she had been recruited for a secret operation by Mossad. She talked of following the mysterious Benamene to Lillehammer, and of sitting next to him in the Skotte's television lounge. She described the identification session in the Karoline. 'I had no idea that a man was going to be killed,' she said.

Lier had emerged once during Gladnikoff's recitation to give the first gist of it to Aanestad. She was dubious and warned Lier to be cautious. But when Lier brought the interrogation to a close after midnight and related the detail of Gladnikoff's account, it was no longer possible to believe that she had made it all up. Aanestad reflected that Norway was a small country several thousand miles from Israel. She asked herself: 'Could this really happen here?'

After breakfast the next morning Lier met Gladnikoff again. This time he was disappointed. A night's sleep had helped restore her composure and she was deeply regretting her talkativeness. She refused point blank to throw further light on her extraordinary story. Raphael and Gehmer were equally tight-lipped. The main hope rested with the man who called himself Ert. At 2:40 that afternoon Inspector Ravlo led Aerbel back to the room on the third floor. Ravlo did not know it yet, but Aerbel was in a very vulnerable state.

A stay in the cells may have been therapeutic for Gladnikoff; for Aerbel, it was agony. It brought back all the terrible memories of the two months he had spent bricked into a cellar in Denmark as a child, hiding from Jew-hunting Nazis. That was why he had asked for his house in Herzliyya to be designed with so few interior doors. That was why, after listening to tales of life in Hamburg jail from Torben Hviid, he once had said: 'I couldn't stand a single night like that.'

When the second interrogation began, Aerbel falteringly repeated his businessman story. Ravlo was sceptical, though not hostile. Then came the long silence, as Aerbel stared out of the window, and Ravlo could feel the sweat on his palms. Finally Aerbel cracked; the subsequent interview lasted eight hours. To occasional prompting by a bemused Inspector Ravlo, Aerbel told everything he knew about the Lillehammer operation, confirming all that Gladnikoff had said, and providing much more, including the emergency telephone number Mossad had given him at the start of the operation. He even gave Ravlo information which later led E-Gruppa detectives to an Israeli diplomat's apartment in Oslo. There they found 'hit team' member Zwi Steinberg and arrested him despite strenuous official protests. For good measure they took from the apartment a communications expert named Michael Dorf, who worked at the Israeli Embassy.

Aerbel also attempted to justify the 'hit team's' mission. He told Ravlo that Israel was at war with terrorists who sought her destruction, and that the Lillehammer operation had been sanctioned at the highest level. Aerbel mentioned that when he saw the cruising police patrol car near the apartment building on the Saturday afternoon he felt sure that the Norwegian authorities were co-operating in the mission. It dawned on Ravlo that Aerbel believed the governments of Norway and Israel had drawn up the Lillehammer plan together, and that he had only to convince Ravlo of this to go free.

It was nearly midnight when Ravlo ended the interview. Aerbel was depressed to learn he would have to spend another night in the cells. Ravlo noticed his dismay and tried to comfort him by saying he was not the only one to have talked. For that solace Aerbel was grateful. It marked the beginning of the dependency Aerbel came to feel for Ravlo, a feeling which Ravlo did his best to encourage and exploit.

Over the next two weeks Ravlo interviewed Aerbel almost every day. The meetings were informal. They ate sandwiches, Ravlo drank coffee, and he saw that Aerbel was provided with the rose-hip tea he favoured. As Ravlo catered to

Aerbel's predilections, Aerbel spilled more and more details of Mossad's operations and the agents who had undertaken them.

On 5 August Aerbel and the five other prisoners were split up. The Norwegians wanted to ensure there could be no mass breakout, or even a break-*in* by an avenging force of Arabs. Aerbel was moved north to Trondheim, an ancient university city 350 miles from Oslo. Ravlo secured for Aerbel a spacious cell in Trondheim prison's modern hospital wing. The cell had hot and cold water, a writing table, walls painted in restful pastel colours, and a fresh vase of flowers each day. The two large windows had no bars and Ravlo even persuaded the prison authorities to leave the cell door open, to ease Aerbel's claustrophobia.

Because of the distance to Trondheim, Ravlo's visits became less frequent. To compensate, Ravlo obtained special permission for Aerbel to keep a typewriter which his sister had delivered. In Ravlo's absences Aerbel wrote him letters, sometimes two or three a day, containing fresh information. By now Aerbel regarded Ravlo as a true friend. Once Ravlo munched an almond from a food parcel Aerbel had received and remarked how tasty it was. Soon a package of almonds arrived for Ravlo from Israel. It was exactly the kind of relationship Ravlo desired. As he rather clinically put it: 'It is a fundamental basis of an inquiry to create an atmosphere of confidence.'

Aerbel continued to provide fascinating details of his work for Mossad. He told Ravlo it began in 1963. He listed the many names he had used, and said that on Mossad's instructions he had reported his Danish passport lost several times in order to obtain new ones. He had given the surplus passports to Mossad.

Aerbel told Ravlo of sailing into ports in the eastern Mediterranean to spy on Arab ships. He made a point of mentioning the presence on the yacht of the lovely, long-legged Sylvia Raphael. Ravlo told Aerbel he was not sure how much he could believe. Such scepticism only encouraged Aerbel to brag more. 'He was always eager to make me believe that he was an important member of Mossad,' says Ravlo. He judged Aerbel 'very naive'.

On 13 August Aerbel asked to see Ravlo again. By then Ravlo was increasingly occupied on other aspects of the inquiry. Although Gladnikoff had given a more or less full account of her part in the operation, her accomplices were proving far less helpful and there were many mysteries for E-Gruppa to resolve. Nevertheless, Ravlo flew up from Oslo on 14 August and went into Aerbel's hospital cell. Aerbel told Ravlo he had remembered one or two more details from Lillehammer, which he now related. Ravlo composed them into a formal statement which Aerbel signed. Aerbel seemed pleased that he had been able to help Ravlo once more, but the inspector was far from satisfied: was that all Aerbel had brought him to Trondheim to hear?

Aerbel genuinely did not have a lot more to say. He had given away most of his secrets. Almost all he had left was the little he knew about Operation Plumbat.

He asked Ravlo if he had told him about the ship. What ship, Ravlo asked?

'I owned the *Scheersberg A*.'

Ravlo asked, so what?

'It carried the uranium to Israel.'

Aerbel immediately realized he had gone too far. When Ravlo probed further, Aerbel would add only that he had not known about the uranium shipment at the time he owned the *Scheersberg A*. Puzzled, Ravlo withdrew. He had made some notes during the conversation, including the words '*Scheersberg A*' and 'uranium' and he typed an account of what Aerbel had said, amounting to 120 words on a single sheet of paper. It was, of course, that brief report which the Norwegian secret service found in the Lillehammer file, and which provided the vital missing piece of the Plumbat puzzle.

Chapter 18
Friends at Court

Zwi Zamir, head of Mossad, had been in Norway to oversee the murder of Salameh. Mossad's leaders have a traditional liking for being on the scene as an important operation nears its consummation, and Zamir travelled to Norway several days before the killing, spending that Saturday in an unpretentious motel on the road forty miles south of Lillehammer. As soon as he heard from the team leader Georg Manner that the 'hit' had been performed, he checked out of the motel and went to Oslo. Manner joined him there. It was not until the first newspaper reports of the murder were published on Monday – there are no Sunday newspapers in Norway – that Mossad knew it had killed the wrong man. Worse, Zamir learned that four of his agents had been arrested.

Ahmed Bouchiki's actual killers, Ingleby and Tamara, had already escaped. The remaining members of the team were still hiding in Oslo, intending to depart in ones and twos. They were now clearly in jeopardy and Zamir instructed them to keep their heads down while he attempted to sort out the mess.

Neither Zamir nor any other Israeli official could openly approach the Norwegians, for to do so would acknowledge that those in custody were Israeli agents. But Israel had always enjoyed friendly relations with Norway, and Zamir thought that a discreet approach might work. A Copenhagen lawyer named Isi Foighel, a prominent member of Denmark's Jewish community, was commissioned to go to Lillehammer. He went by seaplane which landed at the north end of Mjosa Fiord. He went directly to Lillehammer police station and asked to see the prosecuting authorities. He was rather quickly sent away.

If Zamir needed further proof that Norway was not going to overlook the murder it came the following day, when acting on Aerbel's information the E-Gruppa raided the Israeli diplomat's apartment, ignored claims of diplomatic immunity, and arrested Zwi Steinberg and Michael Dorf. When the Israeli Ambassador went to the Norwegian Foreign Ministry to protest this outrage, he was sent away. Israel's chief legal adviser flew in from Tel Aviv to add weight to the protest. The Foreign Ministry remained unmoved.

There could be no doubt that Norway was going to prosecute anyone it could hold culpable for Bouchiki's murder. Zamir and Manner decided the time had come to leave town. They slipped away, as did the remaining members of the 'hit team' who had managed to escape arrest. But Mossad could not let the matter rest there. Events were taking place elsewhere in Europe which made it very clear that at least one of the arrested agents was talking his – or her – head off. Mossad's European network was in danger of being blown wide open.

The revelations in the Lillehammer file were being passed, in a flow of telex messages, from Norway's secret service to its European counterparts. Nowhere were those telexes more eagerly read than at the Paris headquarters of the Direction de la Surveillance du Territoire (DST), the French counter-intelligence service.

The French were heartily sick of the succession of shootings and bombings that the Arab-Israeli conflict had imported to Paris. But the DST had been unable to stop the mayhem or catch those responsible for it. The information that Aerbel was supplying about Mossad's network in France therefore made consummate reading. When Magnhild Aanestad asked for permission to go to Paris to follow up the Lillehammer case she was made most welcome.

The information provided by Aerbel, and gleaned from the possessions of his less talkative companions, led Aanestad, accompanied by two DST men, to a series of Mossad apartments. Aanestad discovered that Mossad referred to the apartments not by their addresses but by the Paris cinemas they were nearest to: Eiffel, Tours and Madeleine, for example. In all she found eight apartments which Mossad had used as accommodation for resident agents or as 'safe houses.' In one of these, she found Mossad's store of sophisticated electronic surveillance equipment.

More significantly, Aanestad discovered ballistic evidence that linked the killing of Ahmed Bouchiki with the shootings of Basil Kubaissi in Paris on 6 April 1973, and of Wadal Zwaiter in Rome on 16 October 1972. All three had been shot with .22 calibre pistols with bored barrels whose boring went to the right. In each case the bullets used had been 'long rifle' cartridges made by the German weapons company Rheinisch-Westfaelische Sprengstoff. In each case they were marked with the capital letter R and striated on the bullet end. If Mossad had killed Bouchiki, then here was clear proof that Mossad had been involved in these earlier murders. (So convincing did this evidence appear that the Italian magistrate investigating the Zwaiter killing later issued arrest warrants against all the named members of the Lillehammer team.)

During her visit Aanestad also found interesting background on two of the suspects in custody in Norway. Since his arrest Gehmer had refused to change his story that he was Leslie Orbaum, a schoolteacher from Leeds, England; he had merely been visiting Norway on vacation, and he had nothing to do with Israel. The DST was immediately able to identify 'Orbaum' as the Israeli diplomat who had been based at the Paris Embassy until 1969.

The DST was also able to reveal the true identity of Steinberg, the Mossad agent whose arrest had aroused furious diplomatic protests. A DST agent took one look at a photograph Aanestad had brought and said: 'My God – that's Zipstein.' The French had good reason to remember Victor Zipstein: he had been Admiral Limon's 'chauffeur' on the momentous visit to Cherbourg on Christmas Eve, 1969.

With this quality of information, coupled with Aerbel's detailed confession, Norway had a formidable case against the 'hit team.' It could demonstrate both that the Lillehammer murder had been sanctioned by the Israeli government, and that it was only the latest in a series of official political assassinations. Mossad also knew, from the Paris raids, that the information Norway was acquiring was dynamite. If it emerged in the formal setting of a trial such evidence could wreck Israel's relations with half of Europe and cast Mossad once again in the role of international outlaw. Israel had to do all she could to prevent the information from becoming public. To a very large extent she succeeded.

The man in the hot seat was Hakon Wiker. As Norwegian State Prosecutor for East and South Norway, as well as the city of Oslo, he had arrived in Lillehammer soon after the E-Gruppa, and had followed the investigation closely from that moment. A tall, handsome and sometimes anxious man, he was a former policeman on the Oslo force who had studied law in order to join the police legal department, and had graduated to his post as state prosecutor in 1968. It was now his task to prepare the prosecution case against the six suspects.

Wiker soon found himself under considerable and conflicting pressure. Norway's Prime Minister Trygve Bratteli was asking for daily reports on the progress of the investigation. In addition, Wiker met frequently with Rolf Jahrmann, the head of the E-Gruppa, and with Gunnar Haarstad, boss of Norway's security service. On the one hand Wiker understood the natural police desire to include all evidence about the motives of the accused. On the other, Wiker was aware that the security service would want nothing to harm its cordial relations with Mossad: Haarstad and Zamir knew and liked each other, and the two intelligence outfits had long exchanged information about terrorist organizations and activities.

In the end Wiker adopted a rule of thumb that he said was fair to both sides: 'I put in just the evidence concerning the murder.' In preparing the case, Wiker excluded large amounts of information about Mossad's operations, 'in the interests of international security.'

Even then, some of the evidence 'just concerning the murder' did threaten Mossad. It did not have to worry. When Aerbel and his five co-defendants went on trial in Oslo on 7 January 1974, the prosecution and defence soon reached an extraordinary understanding. Whenever sensitive matters were raised, such as Aanestad's discoveries in Paris, the packed court was cleared. Aerbel remained as voluble in the witness box as he had been during his interrogations. And after he had once again revealed Mossad's emergency telephone number in Tel Aviv the court was closed, and it remained closed until the end of the trial.

Aerbel's admission that the *Scheersberg A* had taken uranium to Israel was simply not mentioned. Wiker had read Ravlo's brief report on the subject but did not know what it meant. The Norwegian secret service men who also read the file did understand its significance. But they never told Wiker. To Mossad's enormous relief Plumbat was one of the secrets that remained buried in the Lillehammer file.

Norwegian justice was merciful.

Dan Aerbel, Abraham Gehmer, and Sylvia Raphael were found guilty of murder and espionage. Gehmer and Raphael were sentenced to five and a half years; Aerbel, to five. The sentences were less than Norway's usual minimum.

Marianne Gladnikoff received two and a half years, Victor Zipstein one year, and Michael Dorf was acquitted. The sentences were academic since all five convicted were given generous parole.

Aerbel was released after nineteen months, proper allowance having been made for his 'medical condition.' He returned to Israel, to his old job at Osem, and to his house with few doors.

Chapter 19
Closing the Stable Door

Nearly a decade passed after Mossad's Operation Plumbat before even the fact that 200 tons of uranium had disappeared became public. The European authorities deliberately covered up 'the loss' for as long as they could because they found the whole episode acutely embarrassing.

To begin with it had been quite a while before Euratom, the European Economic Community's nuclear watchdog, itself suspected that the Plumbat uranium had gone astray. It was seven months before it knew for sure. Even then it had no clear idea, and certainly no proof, as to where the stuff had gone.

The first hint that something was wrong came two months after the *Scheersberg A* had left Antwerp on that cold November morning in 1968. Under Euratom's regulations SAICA, the Italian paint company, had until 15 January 1969 to report that it had received the uranium. But no word came from Milan. There could have been two innocent explanations: either the uranium had been held up in Italian customs or SAICA did not understand the reporting procedure. On 21 January Euratom wrote to SAICA politely reminding the company of its obligations. A copy of the relevant regulation was enclosed. Euratom also wrote to Asmara Chemie in Wiesbaden reminding the German company that it, too, had an obligation to file a report when it received the uranium. Neither letter elicited a reply.

Another two months passed before the watchdog stirred again. There was still no word from SAICA. The matter was brought to the attention of the director of Euratom's Safeguards Division. He was new to the job, and, as it happened, an Italian – Enrico Jacchia, formerly a university professor.

Jacchia was not alarmed by SAICA's silence. He put it down to the inefficiency of the Italian postal service which, he said, would still be struggling to clear the backlog of mail caused by the Christmas holidays. The Plumbat file was returned to the pending tray.

It was reviewed again in mid-April. The continuing lack of a report from SAICA was readily understandable since Italian postal workers were now on strike. Less easy to explain was the absence of any word from Asmara. It should have received the uranium from SAICA – in catalyst form – some time between January and March and it should have reported that fact to Euratom by 15 April. Professor Jacchia decided that the time had come to shake the tree.

In view of the postal strike in Italy, it was necessary for Euratom to make contact with SAICA by telephone. The call to Milan was not rewarding. On the question of whether it had treated the uranium and dispatched it to Germany, SAICA was, as Euratom later said, 'very vague.' It was also 'very vague' as to where the uranium might now be. Indeed, by the end of the call, it was not even clear if SAICA had ever received the goods. Hoping for elucidation, Euratom telephoned Asmara. Again, straightforward questions got 'very vague' answers.

Firm action was obviously called for. In the middle of May, with the Italian postal strike over, Euratom sent brusque letters to SAICA and Asmara, insisting on some information. SAICA did not reply. Asmara did – five weeks later. Its letter, dated 29 June, was very disquieting.

Asmara professed itself to be puzzled by Euratom's persistent inquiries. The company had purchased the uranium 'on the instructions of a client.' The client had said it wanted the uranium shipped to SAICA but had then 'decided differently.' From that point on the client had made its own arrangements. Asmara had ceased to be involved.

Euratom's immediate reply was brief and to the point: Who was Asmara's client?

The answer, when it came, was from an eminent law firm in the West German capital, Bonn, which announced that it had been retained by Asmara. The law firm pointed out that Euratom's charter did not allow – indeed, expressly forbade

– any invasion by the agency of 'commercial confidentiality'. The identity of Asmara's client was, in the law firm's opinion, a 'commercial confidence': it was none of Euratom's business.

Professor Enrico Jacchia and his men in the Safeguards Division were not well-equipped to take on a recalcitrant company and its smart lawyers. The only weapon they had was Euratom's power to demand the return of the uranium. Since Asmara did not have it, and declined to say who did, Jacchia's men were impotent. They did not even have the authority to conduct an investigation. In forming a nuclear policy for the EEC, the member states had decided against giving Euratom any vestige of police powers. The controversial question of enforcing the policy was therefore left to the security forces of each individual country.

In these circumstances Euratom could only continue to make polite inquiries while the official investigation was put into more powerful hands. In July 1969 the EEC's political executive, the European Commission, informed the six member countries of 'the loss' and asked Belgium, West Germany and Italy to institute separate inquiries. They were conducted with varying degrees of enthusiasm.

The security agencies of Belgium and Italy confined themselves exclusively to selfish questions. Their concern was whether the uranium had fallen into subversive hands within their own boundaries. Having respectively assured themselves that the Plumbat material had left Antwerp and not arrived in Genoa, they dropped the case. The Belgians did not even bother to interview Marcel Heynen and Josef Verhulst of the Belgian transport company Ziegler who shared the distinction of having met the crew of the *Scheersberg A*.

The West Germans, as we have seen, were more energetic. In addition to finding and boarding the *Scheersberg A,* BKA agents spent some time inquiring into the background of Asmara and the two Herberts, Schulzen and Scharf. But the BKA had no power to elicit the name of Asmara's client, and it did not find any evidence that the uranium was in Germany. Neither did it suspect that anyone involved in the Plumbat affair had broken German law. In these happy circumstances, the BKA did not care where the uranium had gone.

Euratom still cared. But Professor Jacchia did not even dare ask the official investigators what they had discovered: they would, he says, have accused him of playing at James Bond. All he could do was continue sniffing nervously around the edges of the Plumbat affair in the hope of finding a clue or two. For the role of bloodhound he had chosen one of his staff, a Frenchman named Pierre Bommelle.

With no more authority than a private citizen – and a foreign one at that – Bommelle went to West Germany to visit Asmara. At the beginning of 1969 the company had moved ten miles or so from Wiesbaden to a wooden shack on the side of a country road – very convenient for Herbert Schulzen since it was only two miles from his home at Hettenhain. Bommelle took along his Geiger counter, just in case.

Schulzen was courteous but unrepentant. He still refused to reveal the name of Asmara's client and claimed to have no idea where the uranium might be. Bommelle, Geiger counter at the ready, asked if he might take a look inside Asmara's shack. Certainly, said Schulzen. It was totally empty.

The visit was, however, not an entire waste of time. While Bommelle was at the shack a truck arrived to deliver a special consignment to Asmara – uranium. It was only 212 kilograms and Asmara had not needed Euratom's approval to buy such a tiny amount. Nevertheless, Bommelle made something of a scene: this was insult added to injury. The next day, thinking better of it, Asmara returned the uranium to its West German supplier.

Euratom's last hope of cracking the mystery lay with Francesco Sertorio, the head of SAICA. The Italian company had never bothered to send replies to Euratom's pleas for information, or, if it had, the Italian post office had lost them. In November 1969, a year after the uranium had left Antwerp, Bommelle went to Milan, still carrying his Geiger counter. It was redundant. SAICA's modest offices clearly could not have accommodated anything like 200 tons of uranium. The offices did, however, contain files and Bommelle asked if he might look through them since Sertorio seemed to have difficulty in remembering anything at all about the Plumbat business. Sertorio affably agreed: he was quite unconcerned and did not even look at the Euratom identification which Bommelle dutifully offered.

The files revealed one very relevant letter. It had been written by Schulzen on behalf of Asmara and it was dated 2 December 1968 – Bommelle did not know it, but that same day the uranium had been transferred at sea from the *Scheersberg A*. The letter said that Asmara was regretfully obliged to cancel its contract with SAICA, and the uranium would not now be coming to Italy for treatment.

Bommelle's amazing discovery refreshed Sertorio's memory. Now he remembered, he said, that he had made a deal with Asmara. On cancellation of the contract they had paid very reasonable compensation of £8,500. He had no idea what could have happened to the uranium. He said he hoped it had not fallen into the wrong hands.

Bommelle returned to Brussels, and Euratom admitted defeat. Jacchia, Bommelle's boss, could only wonder at what he called the 'architectural beauty' of the scheme. The European Commission, representing the six member countries, went into secret session and decided to close the Plumbat file. It also decided to cover up the whole messy affair. The European Parliament, made up of representatives from member countries, and the body to which the Commission is supposed to be responsible, was not informed of 'the loss'. As one Commission official later said: 'It would have made our security regulations look a little ridiculous.'

The Norwegians could, of course, have blown the Plumbat affair wide open at the Lillehammer murder trial. Instead, the cover-up remained largely intact until the summer of 1976 when an intoxicating rumour began circulating on Washington's Capitol Hill. Paul Leventhal, then legal counsel to the US Senate's Government Operations Committee, was one of those who heard gossip that Israel had somehow acquired the means to make nuclear weapons. The rumour mill had precious little information to grind: only that Israel had obtained a large amount of uranium, probably from Europe.

How and when remained a mystery which Leventhal became determined to resolve.

Leventhal was passionately concerned about the hazards of nuclear proliferation. At the end of 1976 he left his government job and, with the aid of a grant from the Ford Foundation, set up an office on Massachusetts Avenue in Washington to research and write a book about the inadequacy of nuclear safeguards. Israel's acquisition of uranium would provide powerful support for Leventhal's thesis and he pursued the rumour with vigour. It was not until March 1977, however, that he stumbled across some solid information, through a chance meeting with the Euratom official Felix Oboussier.

Since Oboussier had approved the contract between SGM and Asmara, he naturally knew a good deal about the Plumbat affair. He also knew that the EEC Commission had decided to suppress the episode. Being a frank man, however, Oboussier, who met Leventhal during a business trip to the United States, saw no reason not to chat informally about what he regarded as a fragment of nuclear history. He never dreamed that Leventhal would make public use of the information. But then he underestimated Leventhal's genuine concern about nuclear proliferation and his determination to broadcast and exploit the lessons of Plumbat.

Leventhal chose to break the story in a highly public manner. He repeated some of what Oboussier had told him in a speech to an anti-nuclear conference held in Salzburg, Austria, in late April 1977. To assure its impact, he leaked the story in advance to *The Los Angeles Times.*

Leventhal's revelation lacked detail. For instance, he did not give the date of the operation or the ship's name, the *Scheersberg A.* And some of his information was wrong. He said that after the heist the ship had reappeared with 'a new name and a new registry,' neither of which was true. News reporters alerted by the *LA Times* story were eager for more – anything, really – but Leventhal could not add much except the speculation that 'the material was unloaded in Israel.'

Perhaps to make up for the lack of detail Leventhal retailed to some journalists a bizarre theory that the mastermind behind the Plumbat Affair could have been a French arms dealer and prince, Michel de Bourbon-Parma. Leventhal said he had not had time to check out this lead but was willing to allow journalists to do so providing he was paid if the story turned out to be true. It was not. In Paris, Antony Terry, European editor of *The Sunday Times*, interviewed Prince Michel who denied the allegation. He said that in 1968 he was busy selling American outboard engines to Egypt for French companies. He added that he would not touch uranium because 'it stinks.' He could only guess that he was being confused with one Michel Ipanema de Moreira de Bourbon, another arms dealer and a pretender to his title, who also lived in Paris. When Terry spoke to this Bourbon he would say only that the other Prince Michel was the culprit: 'I have the proof but I cannot reveal it.' Terry left the prince and the pretender to continue their private feud.

Nonetheless, Leventhal's basic revelation at Salzburg was sufficiently explosive to send the staff of the EEC Commission into shock. Besieged by journalists, the bureaucrats in Brussels would say nothing. Mercifully, the weekend intervened to relieve the pressure, but on the following Monday, 2 May, the siege resumed. The Commission was forced to publicly confess its nine-year-old secret.

Even so, the Commission admitted only that 200 tons of uranium had been 'lost' on the high seas somewhere between Antwerp and Genoa in November 1968. When, three days later, at the EEC's weekly press conference, the Commission's spokesman was bombarded with questions, he revealed little more than the identity of the uranium's purchaser, Asmara Chemie.

If the Commission hoped that would be the end of the matter, it reckoned without the aspirations of Enrico Jacchia. He had left his job as Euratom's safeguards director in 1973 after opting for early retirement and a fat pension. Having been in charge of Euratom's informal investigation into the disappearance of the uranium, he reckoned he had a fact or two to offer. The day after the EEC's press conference Jacchia called one of his own in Rome. As he later described it, in the synopsis of his forthcoming book: 'In a conference of the foreign press, 200 journalists and representatives of television from seven countries, I revealed the operational design of the Plumbat Affair.'

That was overstating the case, but Jacchia did provide more information than the Commission had done. In the process he did little to endear himself to his former EEC colleagues, some of whom had not been very fond of him when they worked together. (Neither did they take kindly to the newspaper articles which Jacchia was subsequently commissioned to write, drawing the lessons of Plumbat. As one former Euratom man put it: 'I suppose he's going to tell us how he solved it.')

By now the European Parliament, to which the Commission is supposed to be responsible, was in a state of high dudgeon. Many members believed that the Parliament should have been told about 'the loss.' Guido Brunner, the EEC Commissioner for Energy, was summoned to the next session to explain why it was not.

When he appeared before Parliament, Brunner, a West German, was labouring under the handicap that he had not been Energy Commissioner during the affair and so had no firsthand knowledge of the events. That did not inhibit him from defending the Commission and Euratom. It did, however, lead to some rather curious statements. He studiously ignored the fact that most of those in the know had believed for years that the uranium had gone to Israel, where it was perfectly suited for use in Dimona's reactor. Brunner said that the Plumbat uranium 'cannot be used as easily as some people think to manufacture bombs.' Indeed, warming to his theme, he added that it was impossible to use the Plumbat uranium for military purposes. Smartly switching tack, he explained that Euratom's regulations had since been tightened to prevent any more of this perfectly harmless uranium from going astray. Commissioner Brunner gave the impression that the whole business had been very unimportant and that Euratom should actually be congratulated for even discovering that the heist had taken place. He was, as one member of Parliament later remarked, like a bank president claiming a reward for the discovery that his bank had been robbed.

Brunner's performance raised more questions than it

answered, and, two days later, another EEC Commissioner stepped up to the rostrum in an attempt to sort out the mess. Wilhelm Haferkamp, also a West German, had been Commissioner for Energy in 1968 and 1969, and was therefore directly involved in the Plumbat Affair. His more intimate knowledge, however, did not make him any more enlightening – or reassuring.

At a press conference in Brussels he blandly announced that little could be done to counter any organized plot to divert nuclear materials from the EEC. He apparently thought little of the new regulations which, he said, he had personally proposed in 1970: they had not been adopted until 1976, Haferkamp said, and then only in 'a toned down form.'

The reporters at the press conference were curious to know, if, at the time of Plumbat, Euratom had possessed any watchdog powers at all. Certainly, Haferkamp said, it could have withdrawn uranium from any company which carried out irregular practices. Unfortunately, Euratom had been unable to withdraw from Asmara what it had been unable to find. Well then, asked the reporters, what about sanctions against the owners of the *Scheersberg A*? That, too, had been impossible, Haferkamp regretted. He said that after leaving Antwerp, the ship had disappeared, had been sold en route and was eventually found outside EEC waters.

That was, of course, sheer nonsense. The *Scheersberg A* changed hands long before it left Antwerp on its momentous voyage. As for 'disappearing,' after leaving Antwerp and making its rendezvous the ship did little else but ply its trade within EEC waters for almost the whole of 1969. Euratom could have found that out from Lloyd's of London for the price of a phone call.

Many members of the European Parliament were bitterly frustrated by their failure to obtain any serious answers from the Commission on what they regarded as a profoundly disturbing episode. Some, including British member Tam Dalyell, came to the conclusion that perfidy on the part of one or more EEC member governments lay behind the Commission's determination to continue the cover-up.

The most popular theory both in and out of Parliament

was that the West German government had actively colluded in Israel's plot to acquire uranium for Dimona. In the United States, for example, *Time* magazine claimed that the then German Chancellor Kurt-Georg Kiesinger had assured the Israelis 'they would be allowed to disguise their purchase [of uranium] as a private transaction in West Germany.' The West German newsmagazine *Der Spiegel* agreed that the coup was probably stage-managed by the Germans and Israelis although it threw in the extra dimension of 'American connivance'.

In November 1977 *Rolling Stone* magazine said that West Germany had conspired in the plot in return for Israel's expertise in uranium enrichment techniques – and $3.7 million in cash. *Rolling Stone* also claimed that Israel disguised West German involvement by sending commandos to stage a fake hijack of the *Scheersberg A* 'as it idled on the calm seas of the Mediterranean.'

A hijack – this time a real one – was also the theory advanced by a paperback book, *Operation Uranium Ship*, published in the US in February 1978. However, the authors of this account claimed that the plot was originally hatched by Morocco, acting secretly for India. Brave, square-jawed Israeli commandos, supposedly hijacked the *Scheersberg A* in mid-voyage – having, of course, disposed of Egyptian secret agents along the way.

The truth is more banal. As we have shown, the uranium was there for the taking. Because of the weakness of Euratom's safeguards Israel had no need to stage a romantic hijack or ask for help from other governments. The conspiracy theory which accuses West Germany is also damaged by the fact that it was German federal agents who, on government orders, conducted the only vigorous investigation into the Plumbat affair.

There is, moreover, no doubt that the West German government was acutely embarrassed by the involvement of Asmara Chemie, a German company and a defence contractor at that. Very conscious of Germany's dependence on Arab oil, the Bonn government naturally wanted the whole messy business to go away as quickly as possible. When German socialist representatives attempted to raise the matter in the Reichstag they were told that the episode was like 'the snows of yesteryear,'* and should be forgotten. When one of the socialists, Gerhard Flamig, telephoned Chancellor Helmut Schmidt's office to press for information he aroused a less poetic response: *'Herunterspielen'* – play it down. Chancellor Schmidt's chief of security told Flamig that the missing uranium was not weapons grade material and it did not matter that it had disappeared. Flamig was left deeply puzzled: 'What nobody has explained to me is why, if that is the case, did somebody go to so much trouble and expense to acquire it?'

*From the line, 'mais ou sont les neiges d'antan' by the fifteenth century French poet Francois Villon.

Chapter 20
Consequences

From the Plumbat uranium, Israeli scientists at Dimona extracted weapons-grade plutonium which was used to manufacture nuclear warheads. They were designed by Yuval Ne'eman, the man who had been primarily responsible for the computerization of Mossad's files. Israel has not tested her bomb but there are few, if any, nuclear physicists who doubt that one designed by Ne'eman would work.

Of all the scientists Israel has produced, Ne'eman has the most brilliant international reputation. He is known as The Brain, and his contributions to scientific research have been enormous. He has served as Director of the Center for Particle Theory in Texas and he was for a time visiting professor at the illustrious California Institute of Technology in Pasadena. In 1969 he became the first non-American to win the Albert Einstein Prize, which was awarded for his contribution to physics. What makes these achievements even more remarkable is that it was not until he was well into his thirties that Ne'eman decided to become a physicist – and he studied for that in what was to him a foreign language.

Ne'eman was born in Palestine – in what is now Tel Aviv – but spent most of his early childhood in Egypt where he spoke French and Arabic. Returning to Palestine in the mid-1930's, he learned Hebrew but found no need to develop any proficiency in English. In 1958, however, Israel needed more physicists and Ne'eman, who had long displayed extraordinary mathematical ability, was asked by the Israeli Defence Ministry to take a physics course at Imperial College, London. First, of course, he needed to be able to speak English and so was sent to London as Israel's supposed military attache. Having mastered the language, he enrolled at the university where he obtained a degree in nuclear physics in half the normal time. Back in Israel Ne'eman was deeply involved in the atomic development programme at Dimona and became Professor of Physics at Tel Aviv University: It was there he designed Israel's bomb.

In preparing the design, Ne'eman was able to employ considerable knowledge of other countries' nuclear weapons, which Israel possessed largely because of the foresight of her first president, Chaim Weizmann. Soon after the Jewish state was founded in 1948 he had sent six of the country's most promising young scientists to study nuclear physics in the United States, Britain, Switzerland and the Netherlands. The knowledge they gained had been of deep interest to her then ally France, who was nurturing her own nuclear ambitions, and Israel had been able to make a deal: in return for their knowledge, Israeli scientists were allowed to study the French nuclear programme. And, until relations between the two countries went into decline, Israelis were allowed to observe French atomic bomb tests. (The deal, which was concluded in total secrecy, also allowed Israel to build Dimona. It was modelled on France's EL-3 reactor at Brest in Brittany and much of the equipment was supplied by France and shipped to Israel under the guise of textile machinery. To provide the necessary cover, the French Commissariat a l'Energie Atomique set up a bogus subsidiary company with officies in the Paris suburb of Courbevoie. France also supplied part of Dimona's initial small stock of uranium.)

Naturally, Israel's nuclear weapons programme was, for a while, her best-kept secret. But ultimately a nuclear deterrent can be effective only if those it is meant to deter know about it. And once the first warhead had been assembled it was not long before positive clues about Israel's new weaponry began to emerge.

The US Central Intelligence Agency had maintained a keen interest in Dimona ever since spy plane photographs had revealed that it was not a textile factory. When the CIA learned – through the US State Department, which had been informed by the EEC as 'a matter of courtesy' – of the missing 200 tons of uranium, it ordered its agents in Europe to find out what had happened. EEC officials were later rather scornful of American efforts: 'They couldn't even find the fucking ship,' one said unkindly. In fact, the CIA strongly suspected that Israel had been behind Operation Plumbat and by 1970, was sure. On 7 July, the Agency's director, Richard Helms, reported to a secret session of the Senate Foreign Relations Committee that the Plumbat uranium was being used at Dimona to develop nuclear weapons: if Israel did not yet have the bomb, he said, she was 'seven and a half months pregnant.'

By 1973, Dimona had produced a small arsenal. The CIA put the number of bombs at three, each in the twenty kiloton range. Other estimates said six. Whatever the number* there is no question that Israel by then possessed her ultimate deterrent and that in late 1973 she was driven near to using it.

At 12:05 pm on 6 October 1973 Egypt and Syria launched the fourth Arab-Israeli war. 6 October was Yom Kippur in Israel, the Day of Atonement, the holiest of all Jewish festivals, and Israel was not on her guard. Of more appeal to the Arabs, 6 October that year was also the tenth day of Ramadan, the day in 624 AD when the Prophet Mohammed began preparations for the Battle of Badr, the first victory in the long campaign that led to his triumphant entry into Mecca. The Arab attack was codenamed Operation Badr and at first it was devastating: Syria attacked from the north, across the Golan Heights, and Egypt from the south, across the Suez Canal. In five days of savage fighting Israel managed to turn the Syrian tide in the north. But she did not have the resources to fight an all-out battle on two fronts and Egypt,

*Current evidence suggests that the estimate of six may have been nearer the mark. In 1978 a CIA document was made public under the terms of the US Freedom of Information Act which said that prior to Operation Plumbat, Israel had obtained 206 pounds of highly enriched uranium from a privately owned nuclear fuel fabrication plant in Apollo, Pennsylvania. That enriched uranium could, at least in theory, have been used by Israel to produce the fissile core of a nuclear weapon. The CIA report was dated May 1974 and stamped 'Secret, No Foreign Disclosure.' However, no firm evidence has emerged to support the allegation. And in February 1978 the US Nuclear Regulatory Commission released a 550-page report in which it claimed that the CIA's theory amounted to no more than 'circumstantial evidence and much colour.'

having secured a massive bridgehead across the Suez Canal, occupied large chunks of the Sinai Desert. Israel was in deep trouble. Her losses of aircraft and tanks were considerable and she had less than four day's stocks of tank ammunition. By 11 October Israel had reluctantly accepted a US proposal that she agree to a ceasefire – without insisting on Egypt's withdrawal to pre-war lines. President Sadat of Egypt rejected the proposal. It was an epic miscalculation.

The pressure of American public opinion alone would probably have forced President Richard Nixon to agree to a massive airlift of supplies to Israel. But there was a more deadly reason why the US could not risk any other course: on the information he had received from the CIA, Secretary of State Henry Kissinger advised that Israel, facing defeat, was near the point of resorting to nuclear weapons.

The CIA was not alone in that fear. Jordan had stayed out of the Yom Kippur war but she watched the events with anxiety and her very efficient secret service reported that Israel had taken the penultimate step towards Armageddon. According to the Jordanians, Israel had equipped surface-to-surface missiles with nuclear warheads and during the first few days of battle, those warheads had been armed.

Israel had always denied that she possessed long-range missiles but the Americans knew that was untrue. During the heyday of French-Israeli relations, the Marcel Dassault company – manufacturer of the Mirage fighter – developed for Israel a two-stage solid-fuel rocket designated the MD-660. When French cordiality cooled, Israel Aircraft Industries Ltd took over development and manufacture of the rocket, code-named Jericho. Very little was known about Jericho except that it could carry a nuclear warhead and that it had a range of 280 miles – sufficient to reach Amman, Cairo or Damascus.

In knowledge of that awful secret, President Nixon authorized the airlift of US weapons and ammunition to Israel. It began on 14 October. With those new supplies, and with a great deal of skill and courage, Israel launched a devastating counter attack. In a week it was President Sadat who was begging for a ceasefire. Egypt and Syria had lost 452 aircraft and two thousand tanks. Israel had crossed the Suez Canal and held a sizeable portion of Egypt. In the

Sinai desert twenty thousand men of the Egyptian Third Army were trapped, cut off from supplies and at Israel's mercy.

Israel could scarcely consider that the Yom Kippur war ended in victory. For the first time the Arab armies were not routed and in the course of the war Israel lost 800 tanks and 115 aircraft. The human cost was appalling: about 2,500 Israelis and 16,000 Arabs were killed. But Israel survived, providing at least short term justification for her decision to take up the nuclear option: if atomic weapons did not deter the Arabs, they certainly deterred the United States government from any thought of abandoning its commitment to Israel.

The question that remains is whether, in the long term, the fruits of Operation Plumbat will bring catastrophe. Can peace in the Middle East ever be built on so explosive a foundation? Will Israel's deterrent one day fail to deter her enemies or coerce her allies, and will she then use it?

Without an Arab-Israeli settlement there seems little doubt that sooner or later Israel will face a terrifying choice. The situation was summed up in an article which appeared in *The Jerusalem Post* on 25 April 1976. It was written by Ephraim Kishon and is, of course, no more than one man's view. Nevertheless, its awful logic seems compelling:

> From time to time the US Administration wonders [in] all innocence why we're so greedy. From time to time it plays dumb and pretends not to know of this tragic situation where three million weary Jews who've just begun building their home in the desert are being forced to maintain a huge military force to defend themselves against a hundred-million millionaires building up an army of NATO size. The US administration acts as if it had no idea that nearly half our Gross National Product lies under wraps in our military emergency stores, and that if it weren't for this back-breaking burden we wouldn't be standing like beggars at their door.
>
> All this generous American assistance, even when it's called economic, goes directly or indirectly to sustain a losing arms race. All the parties involved have an interest

in this race, each for his own reasons – except Israel who can never win it. To be sure, Israel won't be defeated in battle: it'll collapse – economically and socially – under the fearful load of endless arms purchases....

It's a fully planned vicious circle: when the Arabs have 10,000 tanks, we'll need at least 6,000; when they have 20,000, we'll need 12,000 – and so on *ad infinitum*. Interim agreements or not, the race will go on, and our total dependence on the US.

And this total dependence will mean total retreat to the 1967 frontiers and the sticking of a Palestinian State in our throat, *without peace*....

Our one and only alternative to our gradual destruction by arms race is to develop a nuclear deterrent of our own. [Officially, Israel still denies she has nuclear weapons.] It's our single chance for telling our many enemies and our one friend: that's it, we're not playing any more, we refuse to go on running for ever in the circles you've drawn for us. We want no more of your arms, we want a sophisticated educational system.

Sooner or later we'll have to say it out loud. Sooner or later we'll have to announce: if any Arab army crosses this green line we reserve the right to use atomic weapons, and if it crosses the red line we'll drop the bomb automatically, even if this whole country is blown up by nuclear retaliation. You don't believe it? Try us!...

While the consequences of Operation Plumbat remain to be seen, there is no doubt of the value – from Israel's viewpoint – of Operation Cherbourg.

In early 1970, as soon as the fuss over the liberation of the five gunboats had died down, Israel abandoned the pretence that they were to be used for oil exploration. Each was armed with a 76 mm gun and between five and eight Gabriel missiles. They then joined the other seven Cherbourg gunboats to form the elite squadron of the Israeli Navy. Until that moment the Navy had always been the poor relation of Israel's armed services. Once it had teeth it soon proved its worth. In the Yom Kippur War, while the Israeli Army and Air Force came near to defeat, the Navy was utterly victorious.

The Gabriel missile had less than half the range of the Soviet-made Styx missile* with which both the Egyptian and the Syrian navies were armed. But in maneouverability and speed the gunboats were more than a match for anything in the Arab fleets. In every engagement during the 1973 war the Israeli boats got within range of the enemy without being hit.

The war at sea begun at 10 o'clock on the night of 6 October, about eight hours after the start of the war on land. Five miles from the Syrian port of Latakia, five Israeli gunboats ambushed a Syrian torpedo boat and sank it with gunfire. A couple of hours later, in the first ever missile-to-missile confrontation at sea, the Syrian Navy lost a minesweeper and two Russian-made gunboats, all destroyed by Gabriels. A third Egyptian missile boat ran aground and was devastated with gunfire.

Two nights later six Israeli gunboats clashed with four Egyptian missile boats near Port Said and sunk three of them. On 10 October two Syrian missile boats were sunk while still in harbour at Latakia and the next night at Tartus two more suffered a similar fate. After that no Arab naval vessel put to sea in the Mediterranean and the Israeli gunboats had the run of the Egyptian and Syrian coastlines. Day after day their 76 mm guns pounded enemy oil installations, radar stations, military complexes and supply depots. The damage to Israel's fleet was limited to a small hole in the bow of one of the gunboats, caused by a lump of shrapnel.

The Lillehammer debacle did not entirely end Israel's international hunt for terrorists, nor did it temper Mossad's taste for adventure.

On 10 August 1973, Dr George Habash, leader of the extremist Popular Front for the Liberation of Palestine, was due to fly from Beirut to Baghdad, capital of Iraq. But the Iraq Airways airplane he was to take was delayed in London and a substitute aircraft was hired from Middle East Airlines. During the delay Habash, who suffered from a heart condi-

*In 1973 the Gabriel had a range of twelve and a half miles compared to the Styx's twenty-nine miles. Later versions of the Gabriel were given a range of about twenty-two miles.

tion, began to feel unwell and his aides persuaded him to postpone his journey. That decision saved his life.

Although Habash's movements are always clothed in secrecy, Mossad had learned of his proposed trip to Baghdad. When the MEA airliner, a Caravelle, eventually took off from Beirut it was intercepted by two Israeli Phantom fighters. Under threat of being shot down, the pilot was forced to fly to Israel. The Caravelle landed at a military base in northern Israel and was boarded by troops who spent two hours questioning the eighty-three passengers and crew. Although disappointed about the failure to capture Habash Israel's leaders were unabashed at the international outcry over this act of air piracy. Moshe Dayan appeared on television to say that in the war against terrorism, anything goes.

There has been one further demonstration that such Israeli policies remain unchanged although this time the authorities tried strenuously to keep it secret.

On 18 January 1976 an El Al Boeing 707 with 110 people on board was bound for Nairobi Airport in Kenya. An hour before it was due to arrive Kenyan police made a routine inspection of the airport perimeter and came across three Arabs crouching by the fence. On the grass beside them were two SAM-7 missile launchers, loaded and ready to be fired. The three Arabs were arrested.

The Soviet-made SAM-7 is fired from the shoulder like a bazooka, but with far more deadly effect. The missile, which carries a high-explosive warhead, and has a range of almost a mile, is guided by a heat-seeking device. Fired at a jet airplane, it would inexorably seek out the hot exhaust from one of the engines and explode on impact. The Arabs' missiles would have blown the El Al airliner to pieces.

Three days later a young West German couple, Thomas Reuter and Brigitte Schulz, both aged twenty-three, flew into Nairobi. The Kenyan police believed them to be members of the German Baader-Meinhof terrorist gang. They also believed that the attack on the El Al airplane had been jointly planned by German and Palestinian terrorists and that Reuter and Schulz had been sent to Nairobi to find out what had gone wrong. The couple were arrested as soon as they reached the airport terminal. They joined the three Arabs in custody. And then all five disappeared.

In response to inquiries from Brigitte Schulz's parents and the West German government, Kenya denied any knowledge of what had happened to the couple. The first clue to their fate came in June 1976 when German and Palestinian terrorists hijacked an Air France jet and forced the pilot to fly to Entebbe. The daring and celebrated raid by Israeli commandos put an end to that episode but not before the hijackers had listed their demands which included 'the release of the Nairobi five from imperialist jails in Kenya.' The evidence that Schulz and Reuter had been involved in the Nairobi plot led to the West German government making fresh demands of Kenya for information about the couple. Kenya categorically denied that she was holding *any* foreign terrorists. For the first time suspicion dawned that Schulz and Reuter and the three Arabs might somehow have been taken to Israel. The Israeli government denied it.

In December 1976 Schulz's parents hired an Israeli-born lawyer named Lea Tsemel to find out if the government's denials were true. Tsemel is a formidable opponent of the Israeli authorities. She regularly represents Arabs who have been arrested in Israel or the Occupied Territories for 'security' offences and she defends Jews who refuse to be drafted into the army. In her attempts to find the young German couple she pursued the police; Shin Beth, the internal security service; and the Ministry of Defence without pause. Finally, in late February 1977, with the help of the Israeli courts, Tsemel won an admission that the 'Nairobi five' had been taken out of Kenya by Mossad. On 15 March Israel officially informed the West German government that Schulz and Reuter were in custody in Israel. They had been there for fourteen months.

In the autumn of 1977, the 'Nairobi five' went on trial behind closed doors at a military base somewhere in Israel. Tsemel was refused permission to represent any of the defendants on the grounds that she was a 'security risk'. Understandably, Israel was anxious to keep secret the methods by which Mossad was able to spirit the five out of Kenyan custody, out of Kenya itself and back to Israel.

The killings and the reprisals go on.

On 3 January 1977 Mahmoud Saleh, a former PLO official, was shot dead in Paris outside the Arab library he directed at 2 Rue Saint-Victor. On 7 December 1977, David Holden, chief foreign correspondent of *The Sunday Times* was assassinated in Cairo when he arrived to cover the Sadat-Begin peace initiative. On 31 December 1977, two Syrian diplomats were blown to pieces in London by a car bomb. On 4 January 1978, Said Hammami, the PLO's representative in London, was gunned down in his office in Green Street.

There seems little reason to believe that the war of terror will end.

Epilogue

The players who were brought together to make Operation Plumbat possible treat the events of 1968 gingerly, if at all.

The government of Israel is not prepared to discuss the Plumbat affair, contenting itself with a bland statement denying the event 'in respect to its relation to Israel.' As for the country's possession of the bomb, that has always been and still is denied. In January 1978 the press spokesman at the Israeli Embassy in Washington DC said, 'There is no proof whatsoever. I can only repeat to you that Israel is not a nuclear power and that Israel will not be the first country in the Middle East to introduce nuclear arms into the region.' Dan Aerbel, still living in Israel, also denies involvement in the Plumbat Affair and particularly his confession to Inspector Ravlo: 'The Norwegians,' he says, 'are full of fantasy.'

The *Scheersberg A* is still in business. In May 1976 she was sold yet again, this time to a Cypriot Shipping company which renamed her the *Kerkyra*.

In Italy, Francesco Sertorio of SAICA did, when the story first broke, attempt to explain his role, but his stories were confusing. Interviewed by freelance journalist Dalbert Hallenstein on 6 May 1977, he said he had expected the uranium to arrive in Genoa by ship. However, when Hallenstein spoke to him again later the same day he said he had not known that Genoa was the port or even if the uranium was supposed to arrive in Italy by ship at all. There were other contradictions and Hallenstein says that Sertorio was 'very vague.' Unfortunately, it is now impossible to resolve these contradictions because Sertorio died of a heart attack in June 1977. His family are in no doubt that it was the stress of the Plumbat affair that killed him. His son Stefano says, 'The affair was a dirty business. My father was exploited as a front.'

SGM continues to prosper in Brussels and has long since got rid of the rest of the uranium mountain at Olen. In 1977 there was, in Europe, a severe shortage of uranium oxide and Denis Dewez was complaining that he could not fill an order for one kilogram, let alone 200 tons.

Asmara Chemie is no more. It went bankrupt in 1974. An official liquidator immediately went to Asmara's wooden shack near Hettenhain to take possession of the company's records, contained in a steel filing cabinet. Before he could examine them, someone got into the building and opened the cabinet with a blowtorch. The liquidator has no idea what, if anything, was removed. He has found no records concerning the Plumbat affair.

Fortunately, Asmara's bankruptcy did not inflict serious financial injury on its principals. Herbert Scharf retired, handing over ownership of his surviving company Scharf Chemie to Wilhelm Bargon. Bargon says that he knows nothing about any uranium and is certain that the transaction did not go through Asmara's books. 'I would have known,' he says. Scharf himself insists that he, too, knows nothing about the uranium transaction. Ill with a heart condition, he refuses to answer any questions. He simply says: 'Talk to Schulzen.'

Herbert Schulzen is very happy to talk, but not about Plumbat. He now owns his own company, Kolloid Chemie, with smart new premises near Wiesbaden, and he still lives in a pleasant house and drives expensive Citroens. Schulzen will entertain visiting journalists beside his swimming pool with glasses of Coca-Cola. The pool is equipped with a machine for making waves which are the only ripples Schulzen wants to create. When we asked him about his part in the Plumbat Affair he said: 'I'd love to tell you the whole story, but higher political considerations prevent me from doing so.'

Bibliography

Leonard Beaton and John Maddox, *The Spread of Nuclear Weapons*. Chatto and Windus, for the Institute of Strategic Studies, London, 1962.

Bruce Booty, *Guide to Nuclear Physics*. Peter Peregrinus, London, 1972.

Moshe Dayan, *Story of My Life*. Weidenfeld and Nicolson, London, 1976.

Richard Deacon, *The Israeli Secret Service*. Hamish Hamilton, London, 1977.

Halvor Elvik and Mentz Tor Amundsen, *Da Mossad kom til Norge*. Gyldendal Norsk Forlag, Oslo, 1974.

Steve Eytan, *L'Oeil de Tel Aviv*. Editions et Publications Premieres, Paris.

John Francis and Paul Abrecht, Eds., *Facing Up to Nuclear Power*. Saint Andrew Press, Edinburgh, 1976

Nicholas Fraser, Philip Jacobson, Mark Ottaway, Lewis Chester, *Aristotle Onassis*. J. B. Lippincott, Philadelphia, 1977.

Reinhard Gehlen, *The Gehlen Memoirs*. Collins, London, 1972.

Eric Gerdan, *Dossier A... Comme Armes*. Editions Alain Moreau, Paris, 1975.

Isser Harel, *The House on Garibaldi Street*. André Deutsch, London, 1975.

Per Oyvind Heradstveit, *De Medskydilge*. H. Aschehoug (W. Nygaard), Oslo, 1974.

John Hersey, *Hiroshima*. Penguin, London, 1946.

David Hirst, *The Gun and the Olive Branch*. Faber and Faber, London, 1977.

Heinz Hohne and Hermann Zolling, *The General Was a Spy*. Pan, London, 1973. Hardback first published by Martin Secker and Warburg, London, 1972, as *Network*.

The Insight Team of the London *Sunday Times*, *The Yom*

Kippur War. André Deutsch, London, 1974.

Fuad Jabber, *Israel and Nuclear Weapons*. Chatto and Windus, for the International Institute for Strategic Studies, London, 1971.

Robert Jungk, *Children of the Ashes*. Heinemann, London, 1961. First published as *Strahlen aus der Asche*, Alfred Scherz Verlag, Bern and Stuttgart, 1959.

Golda Meir, *My Life*. G. P. Putnam's Sons, New York, 1976.

Henry R. Nau, *National Politics and International Technology – Nuclear Reactor Development in Western Europe*. John Hopkins University Press, Baltimore and London, 1974.

Harold L. Nieburg, *Nuclear Secrecy and Foreign Policy*. Public Affairs Press, Washington, 1964.

Walter C. Patterson, *Nuclear Power*. Penguin, London, 1976.

Shimon Peres, *David's Sling – The Arming of Israel*. Weidenfeld and Nicolson, London, 1970.

Janusz Pieckalkiewicz, *Israel a Le Bras Long*. Jacques Grancheur, Paris, 1977.

C. Richardson, *Uranium Trail East*. Bachman & Turner, London, 1977.

Terence Robertson, *Crisis – The Inside Story of the Suez Conspiracy*. Hutchinson, New York, 1964.

Ze'ev Schiff and Eytan Haber, *Milon Le'bitachon Israel*. Modan, Tel Aviv, 1976.

Colin Smith, *Carlos, Portrait of a Terrorist*. André Deutsch, London, 1976.

Tom Stonier, *Nuclear Disaster*. World, New York, 1963.

David Tinnin, *Hit Team*. Little, Brown, Boston, 1976.

Mason Willrich and Theodore Taylor, *Nuclear Theft: Risks and Safeguards*. Ballinger, Cambridge, Massachusetts, 1974.

Index